I0549527

MICHAEL EDWARD GADDIS

THE LION'S
PROPHECY

THE FIRST BATTLE OF
ARMAGEDDON

This book is a work of fiction. All of the characters, organizations and events portrayed in this fictional novel are either inventions of the author's imagination or are used fictitiously.

THE LION'S PROPHECY
Copyright © 2010 by Michael Edward Gaddis
U.S. Copyright: TXu-1-629-462
All Rights Reserved.

ISBN: 9780988579019
Library of Congress Control Number: 2013933893
Gaddis Laboratories, LLC Defiance, Missouri

For my wife, Kathleen, my beautiful daughters, Alexandra and Samantha, and my two precious granddaughters, Madison and Sydney, I wish for them to live out their lives in peace in this paradise we call the United States of America.

Table of Contents

Synopsis:

Col. Maria Olsen, decorated veteran of the Afghanistan and Iranian campaigns and the general's onetime companion, will be compelled to leave her spiritual retreat and join him on this, his final mission. Together, they will face hardship and mortal danger as Gen. Scofield leads his army to a climactic confrontation with his nation's enemies. Before joining him in death, Maria's last act of devotion will be to seek revenge against his mortal enemy, a mysterious Saudi Prince known to them through the general's Angelic prophecy."

The Lion's Prophecy opens with the nuclear destruction of Washington DC and New York City, and the deaths in those blasts of our nation's most prominent political and financial leaders. Only one man alive has the technological ability and the moral courage to rally the country, rout its enemies, and save the United States from destruction: Gen. Michael Scofield, "The Lion." Will he unleash World War III with his menacing display of destructive power? Will he "stand down" at war's end, or dictate terms to the country's new leadership? Will his secret research into genetic design change the very nature of mankind? These are the questions asked in *The Lion's Prophecy.* With its global action and life-altering technologies, *The Lion's Prophecy* is the ultimate novel for fans of futuristic techno-thrillers. Prophecy also provides food for thought for readers in search of an alternative spiritual zeitgeist. The saga explores the impact of technology on war, and the impact of both on the material and spiritual existence of mankind.

The Lion's Prophecy can be read on different levels—as a fast-moving adventure story, as a cautionary tale about the subduction of our humanity by genetic engineering and other technological marvels, or as a spiritual journey in which the protagonists confront their inner demons and triumph over them.

CHAPTER ONE

DECLARATION OF WAR

AT HOME

"Papaw, she doesn't go over there. She goes in the castle with the other ponies." Gen. Michael Scofield was playing with his granddaughter Tiffany in the main house of his thousand-acre ranch in the Shenandoah foothills of Virginia.

"Sorry, Honey, I should've known. Here…you put her where she belongs."

"You're so helpless, Papaw…" Tiffany said seriously.

Scofield smiled. *Grandkids are the greatest gift*, he thought.

"Michael, come here," Scofield's wife, Sarah, looked alarmed. "Now!"

"What is it? I'm playing with Tiff."

Sarah's expression said it all.

Scofield, holding Tiffany's hand, walked briskly into the great room.

"Papaw, that looks like a mushroom."

Scofield looked down at Tiffany than at the TV, "Sarah, I'm going to get dressed. Call Tommy Marks and tell him to arrange a chopper."

1

The general was putting on his field uniform when Sarah walked into the master bedroom.

"Tommy said he already had a bird in the air. It'll be here in less than twenty minutes."

"Thanks." Scofield looked down and saw Tiffany crying.

"What is it, baby? Why are you so sad?" The general said gently, calming himself with the task of comforting his grandchild.

"Why is everyone crying, Papaw?"

Scofield bent down to hold Tiffany. He looked into her eyes and held her teary-eyed gaze for a long moment, then hugged her tightly.

"They're crying for Papaw because he has to go away. I want you to know that I love you very much; I love your Mamaw and your Mommy too. They will watch over you and protect you." Sarah and Mike's daughter Alison, Tiffany's mother, both came over to hug Mike and Tiffany.

Scofield heard the helicopter in the distance. He pulled away from his family reluctantly. They all stared at him. *Twenty minutes my ass, Tommy...* Scofield thought as he started to walk toward the French doors leading to the back lawn. *Never enough time...*

TWENTY MINUTES EARLIER...

"**M**orning, Madge. What ya up to?" Natalie was Marjorie Hampton's roommate; they had been friends since grade school and had grown up together in the Virginia suburbs outside Washington, DC.

"Not much Nat, have you been outside? It's beautiful today."

"Yeah, John called yesterday and said it was supposed to be warm today. He thought we should go down to the mall and hang at the museums."

"Big spender. When you going to get yourself a sugar daddy, babe?"

"I would, but John is way too beautiful. Maybe I'll just hold off a bit...at least until he starts getting a gut or something." They both smiled.

"I love these early spring days when it warms up—you get a taste of summer. Got to take advantage of them," Marjorie said.

Natalie smiled at Madge. Natalie had just crawled out of bed and was still in her pajamas.

"Maybe the cherry blossoms will come early this year. Washington is so beautiful in the spring. I'm going to go outside and enjoy the morning while I have time; I have to go into town to meet Mom for lunch later."

"Okay, Madge, I'm heading back to bed."

Marjorie shook her head, amazed by Natalie's frequent lazy mornings. Natalie beamed a shameless smile in return.

It was indeed an unusually warm day for this time of year. Following her whim, Marjorie stepped outside her Falls Church home to soak in the warmth of the sun before her day was lost to other tasks. Just a few moments of respite, she thought, before the long drive into the city to meet her mother. She walked out of her house and stopped by the garden's retaining wall. Facing south, she looked at the bare trees that would soon bloom as winter gave way to spring. She smiled and took in a deep breath as she enveloped herself in the unusually warm southern wind. It was 9:43 a.m.

The flash happened so fast and unexpectedly that she could not grasp what it was before it reached her. Her perception of time slowed, dreamlike. She could no longer see from her left eye, but, as yet, it was no more than a weird feeling that something was amiss. She became aware of an intense burning sensation on the left side of her face, and the unnatural light surrounding her—a light too bright to understand, brighter than anything she had ever experienced. *What on earth?* Instinctively, she looked toward Washington.

A deep wave of fear washed over her, nauseating her, compelling her to bend and retch. The brick wall, her makeshift foxhole, had kept Marjorie from being burned more extensively during the blast's initial flash and now protected her from the direct effects of the blast wave. She felt the multiple sensations of searing pain from her face, the deep fear consuming her, the uncontrollable retching, and the roar coming at her that was louder than anything she had ever experienced.

The roar's malevolent source quickly reached her. The blast wave's impact tugged violently at her body, first battering her down then trying to suck her into its vortex. Wind, heat, and debris flew over her head as she hung on for life.

Natalie rushed outside, knocking aside debris as she clawed free of her damaged home. She couldn't see Marjorie through the downed trees and debris blown into their yard from nearby homes.

"Madge, where are you? What happened?"

Natalie spotted movement beside the garden wall, and then recognized the back of Marjorie's head.

"I'm okay," said Madge.

"Thank God."

Marjorie gripped the wall and struggled to rise. When she turned to face her friend, Natalie fainted at the sight.

AT BROAD AND WALL STREETS

It was warm in the van, the heat made worse by the anxiety of its occupants. "We should pray, brother. The Washington bomb should've detonated. We only have twenty minutes left."

"You are right, but I'm worried. The news of the first explosion should be spreading as planned, but they might start checking for anything suspicious, and we are an obvious target in this van on this corner."

"Nothing can stop us now, brother! Say your prayers, and make ready for heaven." As Mohamed spoke those words, a sharp thwack hit the side of the van. Startled, the four men took aggressive stances, crouching, ready to fire their weapons to defend their precious cargo to the death.

It was a New York City policeman; he looked into the van and motioned with his club for the van to move along. In a split second, all five men understood. It was 9:47 a.m.

Mark Harding was at the Fox News desk in Manhattan when he was told that the DC feeds had gone blank. Suspecting a technical glitch, he continued reading the news from his teleprompter. He was good at seamlessly multitasking his many visual, audio, and textual inputs. Suddenly, he paused to announce breaking news. As he spoke, the headline crawled along the bottom of the screen: *A large explosion has occurred in Washington....* Viewers didn't see the rest of the message. Their screens went blank or turned to snow. Mark Harding gasped as he read the teleprompter, his words lagging behind the message on the screen—a message he would not finish.

Mohamed exclaimed, "Allahu Akbar," then picked up the trigger to detonate the weapon. The policeman's eyes locked with Mohamed's, sweat beading on each of their foreheads.

The five men at the corner of Broad and Wall Streets, Mark Harding, and many millions more died in the next few seconds in the consuming fire of a 250-kiloton nuclear device detonated in the heart of Wall Street.

For those outside the blast zones, information was hard to come by. Most folks, who had been watching the news or other channels emanating from New York, started switching channels in search of one that was still broadcasting. Some channels were working; others were not. Nothing seemed terribly wrong at first, until the viewers happened upon CNN, whose headquarters in Atlanta was spared the carnage of the two bombs.

Frank Martin was CNN's duty anchor. Behind him were two split screens, one showing the mushroom cloud from Washington, the other from New York. The clock on the corner of the high-definition television screen showed 10:01 a.m.

"We don't know the extent of the damage. We do know that emergency crews are heading into DC and New York, but as you can see, the bombs are still expending their energy. We're attempting to ascertain the size and type of bombs that went off, but we are having trouble contacting the Department of Homeland Defense or other government officials. At this time, we have no information as to the safety of the President. We do know that he was scheduled to be in the White House, and that there was a cabinet meeting slated for this morning. Let us take a moment to pray that our President and government officials survived this terrible attack."

Frank bowed his head and let a full minute tick by as he attempted to regain his composure.

His producer gently whispered into his earpiece, "Frank, we need to resume; we have some breaking news."

Frank managed to resume: "This just came in. The Washington, DC, bomb has been estimated to be approximately a 150-kiloton nuclear device. It was detonated just outside the White House. The bomb has leveled DC and has sent superheated air and debris blasting in all directions. The mushroom cloud will reach a height of ten miles, and, as you can see from our live feeds, can be seen from miles away. Unfortunately, the New York bomb was much larger— probably twice the size.

"We've located retired Air Force Gen. Frank Starling who spent his career within the command structure of our strategic nuclear forces during the Cold War and is an expert on nuclear weapons and the aftermath of their explosions. He is joining us via Skype.

"Gen. Starling, welcome. I wish it were under better circumstances."

"Thank you, Frank. A terrible day indeed."

"General, what can you tell us about what is happening in Washington, DC, and New York right now?"

"I don't have good news, Frank. For those unlucky enough to be close to the blast, it was probably a mercifully short experience. Most DC and Manhattan residents were immediately vaporized when the bomb detonated, as were the nearest neighbors in the Virginia suburbs just across the Potomac River and the Maryland suburbs to the north, east, and south of the district. Some aboveground victims probably survived the initial blast, but few will survive the radiation exposure that they've most certainly received. You see, Frank, these bombs weren't delivered by missiles and burst in the air for maximum destructive force. These bombs were detonated at ground level—probably positioned in a delivery truck."

"How do you know that, General, so soon after the blasts?"

"If you look at the live video captures of the explosions, you can see by the progress of the mushroom clouds that they were ground-based."

"I'll trust you on that, General. I don't think I can watch those explosions again. What else? What about those who live in the suburbs?"

"Well, Frank, the people in the ring between three and seven miles from ground zero...their luck ran out today. Many more of them will survive the blast than those near ground zero, but most survivors will have multiple injuries, and almost all will have some radiation exposure. Help will probably not come in time for most of them. The roads, hospitals, and emergency responders needed to help ease their suffering will have been decimated. The interior of this ring is likely already ablaze in a giant firestorm that will be nearly impossible to escape or penetrate. The firestorm at Hiroshima killed more than the initial blast. There will be few survivors. God help them."

"How far away from the blast did you need to be in order to survive?"

"Outside of seven miles, the carnage will extend in a great ring of death and destruction. It will seem more random than it should. As any blast wave spreads, it follows the lay of the land. Manmade and natural obstacles will impede the shock wave, and, thankfully, its violence will be diminished yard-by-yard as it expands. Destruction in this band will be bad, but not complete. The injured will be rescued as emergency services from surrounding communities work their way into the outer segments of the blast zone. The first responders simply can't go deeper than the seven-mile ring until the interior fires die down."

"What about New York?"

"The destruction in New York will be horrific. The death toll will reach millions. The national and global economies that wind their financial web through New York will be shattered and in chaos for the foreseeable future."

"Do we know who did it?"

"I'm not at liberty to say. Sorry."

The general knew instinctively who was responsible, though he didn't know which specific terrorist organization had pulled this off. Starling maintained a stoic expression. He dared not speak his mind, so he only said it to himself under his breath and as a witness to his God, "Islamic Terrorism is now a world power, and it has just launched World War III."

"What was that, General?"

"Oh...nothing, Frank."

AFTERMATH

Natalie walked Madge to a local hospital, only to find utter chaos there: a scene straight from Dante's Inferno. The hospital had sustained considerable damage but was still standing. Doctors and Nurses struggled to help as best they could, but they looked totally overwhelmed to Natalie.

Marjorie was suffering less pain than might have been expected; many of her nerve endings had been fried in the flash burn on her face. Some of her flesh was starting to swell, and necrosis was setting in. Her skin sagged; her face appeared melted by the blast.

Marjorie was in deep shock, losing touch with reality, and profoundly losing hope. The mind can distract the dying from their fear and pain, giving them a sense of serenity.

Marjorie saw her mother by the gate, gesturing to her. She smiled and said weakly, "Hey, Nat, Mom's over by the gate; help me up, I need to talk to her."

Natalie tried to direct her toward the hospital entrance. She knew that Marjorie's mother wasn't there: she lived in Georgetown, and was almost certainly dead from the blast.

At about the same time, a camera crew from a local television station happened by the two women. They turned their camera on Marjorie just as she was looking to the gate to see her dead mother. Marjorie's horribly disfigured, once-beautiful face and her calm countenance captivated the cameraman. He began to weep softly. He focused his camera on this pitiable image, a voyeur in hell compelled to his task.

Natalie pulled Marjorie's hand, frowning protectively at the camera crew. Marjorie would have none of her prodding; she pulled away, walked toward the gate, and sat on a bench there.

"Mom, I'm so glad you came, I was worried about you."

Natalie sat next to her friend and held her hand. Marjorie placed her head on Natalie's shoulder and wept. After a time, Marjorie passed. Natalie didn't speak or sob. She held her gently for a time, then rose from the bench and walked away, unable to cope further with the day's events.

BACK AT CNN

The stress on Frank Martin's face was palpable. He did his best to be professional while repeating the words sent to him by his producers, but when he saw the image of the young girl at the hospital on the screen to his right, his tears began to flow. Seeing this, his producer yelled at her assistant to locate one of the senior correspondents.

Frank wiped at his tears, then shrugged and continued:

"Ladies and gentleman, we have preliminary news about our government's leadership. We've been told that it's likely, given who was at the White House, the United States Capitol and the Pentagon, that the President, Vice President, Speaker of the House, and four of the five serving Joint Chiefs of the Armed Forces all died instantly in the blast. Little trace of the White House above ground remains. Congress was equally ravaged. Since the bomb was set off two weeks prior to Congress's Easter recess, many members were either in or around their offices. It also caught many cabinet members, including the Secretaries of Defense and State, as they arrived for this morning's cabinet meeting at the White House. It looks like this was the perfect storm.

"Another tragedy, one that will take many months for us to fully fathom, is the probable loss of the many treasures in the Library of Congress, the Natural History Museum, the American and Air and Space Museums, and many other special places in the district. Much of the history of the United States has been vaporized…sent into the air to rain down on innocent souls as poisonous confetti."

"We have to get him out of there—he's cracking up." Frank's producer snapped a look at her assistant, "Where is Singleton?"

Martin continued: "We've been told that it was likely done by suicide bombers, and since they exploded their nuclear devices at ground level, this has maximized the radiation and radioactive debris. Vast plumes of debris have been blown into the air and will rain down on the two areas for days. Survivors in Northern Virginia will fare better than their counterparts in the Maryland suburbs. The winds are prevailing from the southwest today and will continue that pattern for at least three more days. The majority of the radiation from the blast will likely be felt by Maryland.

"The winds are also blowing west to east today from New York. If you live on Long Island or in western Maryland, if you can evacuate, do so now.

"The reactions from the rest of the country have varied. We've seen some panic in Los Angeles, Chicago, Houston, and other major cities. People fear they will be next. We know the rest of you are staying glued to your televisions. I know you ache for information; we'll deliver what we know as soon as we know it. Please bear with us."

Jennifer Singleton was brought in to replace Frank Martin at CNN. She was a consummate professional, calm, reassuring, empathetic. She would hold the nation's hand for the next twelve hours as she ran down one tragedy after another.

"Folks, it appears the butcher's bill has come due for our nation's lack of commitment in combating terrorism. In Cairo, Baghdad, Tehran, and in numerous other Islamic cities throughout the world, people have filled the streets. They are—amazingly—celebrating this vicious attack on our country. The Arabs in the street are apparently jubilant that the United States has been attacked, ecstatic that the leaders of our government are dead. The Islamic news outlets are emphasizing how Americans brought this tragedy onto ourselves through our arrogance and our so-called misadventures in Iraq, Afghanistan, Iran, and Israel. My personal view, ladies and gentleman, is that they might well regret this day in the future as much as we regret it today."

During the television coverage, a sad, disfigured girl was shown outside a hospital—part of the background coverage to show the human cost on the ground. The image of the girl was repeated many times for filler while news analysts tried to decipher the meaning and extent of damage from the attack. Marjorie Hampton would become the iconic image of a war as yet unnamed.

DIPLOMACY

The leaders of the Islamic countries were not as jubilant as their citizens. They knew that there would be hell to pay for this attack. The United States, now leaderless in their eyes, might even unleash nuclear war on their countries. They contacted local embassies to convey their condolences and to feel out what the United States might do. They expressed great dismay at what had happened, and at the celebrations that were going on. This wasn't, they assured the diplomats, anything more than the uneducated rabble acting out.

US diplomats in foreign lands had no clue what was going on. Their leaders were dead, and they were getting no reliable information or direction. Communications had been restored through military and intelligence channels and regional commands. They were told that the government was re-forming. Ambassadors around the globe met with their counterparts and improvised as best they could, trying hard to suppress their own feelings of panic.

The exception was Pakistan. The US ambassador to Pakistan, Arthur Robinson, had been instructed to tell the Pakistani leadership that the United States had determined the origin of the bombs and that we knew they had *not* originated from Pakistan.

Ambassador Robinson arrived at the Pakistani foreign minister's office and was surprised to see Gen. Dharker rather than the minister waiting for him.

"Mr. Ambassador, please accept my condolences on this terrible news from America." Dharker said.

It was unusual for the ambassador to be greeted by an army general. Robinson knew Gen. Dharker, and knew he controlled the strategic assets of his country.

"Thank you, General. Where is the minister? I was expecting to see him today."

"He was detained with an internal matter and asked if I could receive you today. Is that a problem?"

"No, no problem. I came to convey assurances from my government that they've discovered the origin of the fissile material, and that they've determined it didn't come from Pakistan. I assume you're relieved to hear that?"

Gen. Dharker was a petty man, brash and accustomed to bullying; subtlety wasn't his strong suit. When he heard what the ambassador said, he nearly broke into nervous laughter.

Robinson noted this and thought, *Well, that confirms it.*

"I'm so glad to hear that, Ambassador. As you might expect, we were worried what America might do, now that so many of its leaders are dead."

"I assure you, we have no hostile intention toward Pakistan."

"Forgive me, Ambassador, but with the President dead, how can we assume you speak with authority on this matter?"

"You may assume what you wish, General. Good day." Robinson pivoted and left the room.

The Pakistani interior minister walked into the room after the ambassador had left.

"Well, Minister, what do you think?" Dharker asked.

"If he didn't know before he came in here, your childish glee certainly told him what he needed to know." The minister glared at the general.

Dharker, unaccustomed to criticism, snapped back, "What can they do? They can't do anything without proof. The Iranians will bear the brunt of the American counterattack, and Pakistan will continue its ruse of being pro-American, and anti-terror. We've cut their balls off."

Ambassador Robinson had been intentionally misinformed by his government. Within hours of the blast, a Nuclear Emergency Support Team (NEST) had been dispatched to DC to measure the radioactive remnants and finger the killers. The surviving military commanders back home in the United States were, at the very minute the ambassador was conveying his message to Pakistan's leaders, formulating a military response to punish Pakistan.

THE TEMPLAR DIVISION

U.S. News and World Report journalist Mark Osborne appeared to the casual observer to be a typically cynical reporter, both in writing style and worldview, but he was a closet romantic, successfully hiding a deeply empathetic view of humanity. This strange combination of cynicism and empathy made Mark a good observer of human events and foibles while keeping his reporting free of sentimentality. Mark was among the press members summoned to Camp Lejeune, North Carolina, by a call from the Marine Corps provisional headquarters, now located there. Many of his colleagues had perished in Washington and New York, leaving him among the last national news reporters left standing.

As the home of the Marine Corps Second Division and the largest Marine base on the East Coast, Camp Lejeune seemed an appropriate place to establish a provisional headquarters following the destruction of Headquarters Marine Corps (HQMC) at Henderson Hall, which had been located in the Navy Annex in Northern Virginia, just across the Potomac River from the Pentagon. Henderson Hall—like most government headquarters—lay in a charred ruin, prominent generals entombed within. The Commandant of the Marine Corps had died at his home in the Marine Barracks, on the outskirts of the District of Columbia.

The press sat about twenty yards back and to the right of a small, raised platform in front of a large parade ground that had no entourage, podium, or decoration of any kind. Mark thought the asceticism odd, but he kept his curiosity in check.

Gen. Michael Benjamin Scofield, a man immediately recognizable to most of the assembled reporters, emerged from a mobile command center behind them to their left and began walking silently to the platform. The gait of the so-called "Lion of Afghanistan," was purposeful, relaxed, and athletic. He was the only person there, the parade ground desolate. The general wore a simple field uniform. His only decoration was his first Medal of Honor, draped around his neck. The ribbon holding the medal was worn and bloodied, as if he had never taken it off, had never left the battlefield. It had an odd effect on Mark; he imagined that Scofield had been transported to this spot directly from the Iranian battlefield seven years in the past—a fantasy made more real by his long absence from the world scene.

Gen. Scofield was probably the only man ever to have worn a Medal of Honor into battle on full display to his enemies. He said he wanted his adversaries to know that if they killed him, they had won a great prize. So far, no one had claimed that prize, though many had tried. The general was an odd duck, to be sure.

The reporters were unusually quiet. The general had created an environment of heavy expectations with this austere setting.

Mark pondered how old the general must be: at least sixty-four or sixty-five. He hadn't thought of Scofield for many years, assuming he had retired and faded away. But now he wore four stars, so he must have stuck around out of the limelight. He looked younger than Mark remembered from the last time he'd seen him—certainly younger than he should seem these many years later. *Given his age and many battle wounds, it's amazing that he can still walk and talk,* Mark thought. Over the years, Scofield had received twenty-seven Purple Hearts. It had to be a record.

In front of the Lion's podium, there were no marines, no guns, and no vehicles. It looked a little odd, but Mark was sure that would change. *We weren't brought here just for a speech to an empty field,* he thought.

While the reporters stood there, cameras ready, music began to swell from the parade grounds, seemingly from nowhere. The sound came from the front of the stage, from the empty parade field. The music was rich, powerful, and serenely beautiful. Mark had never heard music so rich in an open field like this, even at the best concerts. It was impossible not to get caught up in the melodious operatic voice. *Angel music,* Mark thought. *Appropriate for the mood of America today.*

The serenade continued as the television cameras panned around, cameramen and reporters looking for visuals to show their live world audience. But there was nothing to see, except the general standing like a rock looking out over an empty parade field, as music from an unseen source grew louder.

Suddenly, Mark sensed something new in the scene unfolding before him: the sound of marching, feet pounding against earth, muffled but persistent. It would take thousands of feet to produce that sound. *Where are they?* he wondered.

Then Mark glimpsed movement on the parade grounds, movement that could only be described as that of ghostly apparitions: fleeting images, a silhouette, a leg, an arm. Each image phased in and out of view in weird, uncanny ways. The overall effect was eerie.

Mark knew the images were being created and manipulated by a scientist's new alchemy, presumably for emotional impact. It didn't matter that he knew he was being deceived; he was transfixed by the spectacle. Something important was happening here, and he was standing just yards from it. This technology was new—something no one had seen before.

It was clear to Mark that the unveiling of this technology at this time, on this day, was a message that the Lion was sending intentionally. But the demonstration wasn't for the reporters or their viewers. Mark understood that Scofield was speaking directly to his adversaries.

Despite the compelling display in front of him, his mind began to drift. Yesterday's events haunted him. Like the image of that beautiful girl—a rare, perfect beauty easily recognized by the undamaged side of her face. Mark could not forget how she looked. Beauty transformed. The horrid spectacle of her disfigured face was seared into his mind; no matter what image he recalled, hers at last, would stare back at him. Beauty melted into mush like in a scene from a horror movie. Mark couldn't stop replaying that terrible image. The wounded woman showed no evidence of pain or shock—just a heartrending, distant stare. *What was she thinking at that moment that gave her such serenity?* That thought haunted Mark.

Mark snapped back to reality as he heard more footsteps, saw more phantom glimpses of movement. It was obvious to all of the reporters that the troops parading in front of them were invisible. *How the hell are they doing that?*

The music grew louder, transcending hearing and becoming a physical sensation. Mark felt the vibrations of each note inside his body. It was unlike anything he had experienced. *What is that piece?* he wondered. He hadn't heard it before. It was inspiring, sad, and beautiful when played this way—matching his dark mood.

"That's Maria Olsen," said the know-it-all on Mark's right. "I don't recognize the music; it must be something new. I thought she had retired and become a nun." Mark didn't answer; the moment was too perfect. The music ended, slowly transitioning to a soft orchestral background tune that Mark recognized from the movie *Platoon.* The music was sad. That fit his mood too.

The marching sounds stopped. They must be in place.

The general began to speak. Except for the music, softened but still ethereal, it was deathly quiet. The general had no perceptible microphone, yet his voice seemed to come from everywhere, like the music. In fact, the two sounds were perfectly integrated, like a movie, only live and in three dimensions.

The general spoke slowly, in measured tones and with careful enunciation. "Yesterday our nation was most grievously attacked by our Islamic enemies. The destruction of two of our greatest cities is complete, and the effects will be lasting. Great treasures have been lost. This wound will never fully heal," he said, without emotion.

"This attack struck us as deeply as it was meant to; there was no mercy shown or quarter given to the millions of dead: women, children, old and young." Here the general paused briefly.

"There will be a *reckoning*," he said with deep conviction and passion. Then, in a command voice any drill sergeant would envy, the general called the formation to attention. All at once, the field was populated with Marines in battle dress. They appeared from nothing, simultaneously. Many thousands stood in formation, dressed in uniforms Mark had never seen before. The uniforms appeared to be some form of exoskeleton; the Marines were completely clad from head to toe with a thin black layer of material that clung tightly to their muscled bodies. It was truly remarkable, seeing so many appear from nowhere instantaneously.

Good show, General. Impressive indeed, Mark thought.

"Aruuugaah!" the general boomed, yelling the Marine Corp's battle cry.

In unison, with what seemed like a million voices, the field resounded with the deafening response: "Aruuugaah!"

The battle cry and magic tricks achieved their desired effect, touching Mark profoundly. It was scary, powerful, and majestic all at once.

The general commanded his regiments to report their status, and each commander did so in turn. Then the general spoke again. "We go now to Arabia to attack the heart of our enemies. We do not go as peacekeepers, or to spread the democratic faith to the hopeless. We do not go to negotiate or to appease. We land on our enemies' shores to seek *revenge*." He spoke with controlled fury and venomous effect, slowly enunciating each word.

The troops shouted their approval in battle cries. They, like the observers, were caught up in the moment, overwhelmed by the jarring emotions of the previous day's destruction and their own unknowable fates.

Looking up and down the line, the general said, "I shall not return until the heads of these fascist Islamic bastards are impaled upon the standards of my regiments. To this I swear my fidelity. I pledge my life and the lives of my regiments."

Next he addressed his commanders: "Take charge of your units and board your assigned ships. I'll brief you when we are at sea." To the troops, he said: "I'll see the rest of you on the beach. Strength and honor!"

The troops replied in unison, and somberly: "And a good death!"

CHAPTER TWO
SAILING IN HARM'S WAY

EMBARKATION

Mark Osborne was asked to join the embarking force as an embedded reporter for *U.S. News and World Report*. The home office cleared his assignment enthusiastically, and he proceeded to "pack his trash," as the Marines said, and make ready to board the ship. He had less than twelve hours before sailing.

Mark threw his things into two small bags and drove to the naval base in Norfolk, Virginia. The guards at the gate scrutinized his credentials and patted him down before issuing a base pass to drive to his assigned ship.

There were no ships at the docks. They were out to sea, or somehow cloaked. *How is it possible to keep such a thing secret?* he wondered.

Mark hailed a naval officer, who directed him to a gangway a few hundred yards down the pier. The ensign there led him to a gangway that looked like it was attached to open air. Mark could see through the covered gangway into the ship as if he were looking into the mouth of a cave. He popped his head back out of the covered gangway to make sure the ship was still invisible. It was.

Oz was becoming blasé about such magic. He walked into the invisible ship and introduced himself to the Officer of the Deck. The officer greeted Mark warmly and had a young sailor show him to his stateroom. He was surprised to find that he rated a billet in "officer's country," the collection of private and shared staterooms aboard ship where officers lived and worked when not at their duty stations. Mark was glad to have the space, rather than sharing enlisted quarters crammed with Marines.

"So, you work for a newspaper, sir?" The young sailor asked.

"US News and World Report."

Mark's credentials didn't impress the sailor. "Oh. Okay, I have some things I need to show you. Your assigned head is over here," he said, pointing to a door down the passageway. "And the officer's mess is down that passage towards the port side of the ship. Ask anyone, and they will help you find it."

"Thanks."

"You must be pretty important to be bunking up here near the general," the young man said. "The rest of your companions are generals, colonels, or Gen. Scofield's personal bodyguard. I was told to tell you that the general will be addressing his staff later today and that you're invited to attend."

Mark smiled and started to unpack his gear, happy that he was being afforded access to the top echelons of Scofield's staff. *My editors will be ecstatic,* he thought.

MEETING THE GENERAL

"Come in, Mark. Have a seat," the general said casually as he motioned Mark to a chair near the center of the room.

The general's quarters were lavish, as shipboard accommodations went. He had two main rooms: an outer meeting room with an attached private bedroom. He also had his own private head, a true luxury aboard a naval combat vessel. Along the port side of his meeting room was a small office with computers and displays sending round-the-clock updates to the general. The furnishings were rich and well designed.

"Are your accommodations to your liking?"

"Yes, sir, they are," Mark replied, "though I was wondering why I was the only reporter in the flag quarters section. Why me?"

"The guy we wanted is dead," the general answered. "In fact," he added, with a smile, "there were five guys and a gal in front of you, but they're all dead now. You people should learn something about dispersion."

Mark ignored the jibe. "Why do you want someone from the press so close to your headquarters?"

"There are rules that you must agree to," Scofield explained, ignoring his question. "This assignment isn't about counting coup on your peers; it's about understanding and observing."

"I'm not sure I know what you mean, General."

"You'll be a witness to history. Even though I've made sure that every scene will be recorded at a thousand different angles, seeing a thing from your living room isn't the same as knowing a thing by having lived it." The general seemed to ponder his own meaning, as if remembering his own combat experiences.

"I want you to be a witness to what is going to happen. I'll commit significant time to you to help you interpret what you see. I want you to understand what happens at a human level, to give the battle color and empathy."

"That's not my job, General."

"I know, but in time, you'll come to understand what I mean."

"I can't promise anything."

The general reached for two small goblets and filled each with brandy. Handing one to Mark, he sat down facing him and said, "Tell me about yourself, Mark."

As the general listened patiently and attentively, Mark recounted the highlights of his so-so resume, coloring his history with self-deprecating descriptions of his education, assignments, and experiences.

"So, basically, although you're amazingly well qualified, you have been limited in your career because you're a smartass, a little bit lazy, and openly defiant of authority?" the general summarized. "Maybe you have a chip on your shoulder."

"That's an odd comment to make, General," Mark said. Such deeply accurate insights could not have been gleaned from the recitation of his resume. *He must have good spies,* thought Mark.

"What do you want to know about our little operation, Mark?"

"How much time do we have, General?"

He smiled. "Not much. I guess we'd better get started."

"Do we know who exploded the bombs?"

"Yes."

"Who? Where are we headed, and what do you intend to do when we get there?"

"This is classified, so you cannot use this information, but I'll tell you. The fissionable material came from Pakistan with the full knowledge of its government. It was sent to Iran to be made into four bombs, again with that government's full knowledge and approval. Two of the bombs were given to a US-based Hezbollah cell with Al Qaeda ties with specific instructions as to how they were to be used." The general elaborated, in far more detail than Mark would have expected.

"Do you intend to attack Iran and Pakistan, then?"

"I intend to pay a visit to Saudi Arabia to pay my respects to the prophet at Mecca." Scofield said this matter-of-factly, letting his meaning sink in. "But we'll get around to them later, after I lay waste to the holy cities."

"You mean that, General?"

The look on his face gave the answer.

"You laid waste to Iran years ago, General. Now you're going to trash the Holy Cities? Is there any way back after this?"

"Back to where?" the general asked rhetorically. "What's left but to destroy them all? If I am not successful, that will be the world's fate. Either the Islamists renounce their self-destructive war, or the political leadership in our new government will turn the entire Middle East and Indonesia into fused glass."

"So you're trying to save the Islamic countries by attacking them and laying waste to their most treasured cities?" Mark asked acidly.

"Maybe…Maybe."

"Are you going to tell the Saudis that you intend to invade them?"

"I'll leave that to you, Mark."

"So I can tell the world what you have told me?"

"Yes, but leave out the Pakistani and Iranian connection. We'll use that later for a little surprise."

"The whole of the Islamic world will rise up to defend their holy cities, General. You have with you—what, ten thousand men? This looks like madness."

"I'm counting on it. Yes, we have about ten thousand combat troops, and many more in support. We also have a few sailors and airmen along to help us."

With that, he stood and motioned for Mark to do the same. "Time to go, Mark. We'll talk later."

THE REACTION

Mark Osborne broke the news that the United States of America would invade Saudi Arabia and attack its holy cities, Medina and Mecca. The news wasn't well received around the world. The world's media organizations could not mount much of a protest, though, since the American Government had not yet reformed.

If there was new leadership, no one was talking about it. The US Government was operating on many levels; cabinet departments were functioning, state and local police were operational, and the military was functioning at the highest level possible. That had been made clear by the many news stories about Homeland Security, FEMA, and the FBI. Many other agencies were fully engaged in crisis management and recovery. These stories made hourly news headlines, always ending with some agency head saying that the top levels of government were being formed and would be announced soon. After four days, this reassurance was wearing thin.

In truth, as only a few of the nation's power brokers knew, the fight to name the next President of the United States and his many constitutional counter-parts was coming to a head. If the competing parties to this selection didn't remain calm and cooperate, civil war loomed—or, worse, a military coup. There were two camps: the progressives, who believed the nation could persevere and restore their country's greatness without resorting to barbaric violence; and the hardliners, who felt that we had suffered this blow because we had been weak and that we needed to strike a fatal blow to Islam and the nations that nurtured its twisted ideology.

The latter camp was divided between the political right and a group that could best be described as center-left leaning moderates. Given the severity of the attack, centrists were now embracing a tougher approach. The right, coupled with the centrists, constituted a strong political majority in the United States. If this side won—Scofield's plan be damned—the United States was going to nuke the bastards.

The voices of the progressives, mostly drawn from the liberal political ranks, had been muffled by the hardliner's cries for revenge as they stoked their constituents' now open hatred for the Islamic peoples. Because of this pressure, the progressives had hastily green-lighted the nebulous "Scofield Plan," hoping that the engagement of troops would provide solace to the American public and slow the budding hardliner movement. The progressives rationalized that the hardliners would have to wait at least until Scofield finished his mission. Still, such a plan was fraught with risk. If Scofield failed, if he and his men were slaughtered, there would be no stopping the hardliners. Scofield would become the modern-day Custer and give the hardliners their rally cry for nuclear retaliation.

The military was the arbiter of this debate. Each side of the political landscape was wooing different military factions simultaneously. But the military leadership itself was fragmented because of the losses of so many of its senior generals, including the Joint Chiefs. The military leadership might have more quickly repaired itself if it had had the luxury of doing so under stable civilian oversight, but the President and Secretary of Defense were both dead. The four-star commands in the different services were holding their own for the time being.

TECHNOLOGY IS THE KEY

"Well, Mark, you have certainly stirred up a hornet's nest," Mike Scofield observed.

"It's what you wanted, General. I was just doing your dirty work."

"There are people who are accusing you of treason for leaking this information, Mark. You'd better hope I'm not killed, so I can testify at your trial."

"Great, thanks. I don't suppose you could write me a letter to put in my safe?"

"Sorry. Then they would accuse *me* of treason."

"General, will you show me your new toys? How you achieved invisibility?" Mark asked.

"We've achieved much more than invisibility, Mark. The technology we've invented is a quantum shift in warfare and, for that matter, in human engineering. You'll see its effects soon enough. I could give you a little demonstration, if you like. That is, if you have the stomach for it."

"Of course, I'd like to see it. How about a technology primer first, so I'll understand what I am seeing and how it was achieved?"

"You would need quite a bit of time and instruction before the pump would be primed. See the demonstration I've put together for a few visiting politicians, and then we can talk about how we did it."

"Do I have any choice, General?"

"No," he said, smiling. "Read up on the science of nanotechnology."

TROUBLE AT HOME

"General, we need to talk about what's happening at home," said Kentucky Gov. Raymond Jones. "They attacked us at a bad time. Everyone was in DC. We only have seven surviving senators and forty-seven congressmen to rebuild our government. Most of the senior government officials are also dead. We must act decisively."

"We *are* acting decisively."

"What? By attacking with ten thousand men? What possible difference can you make with such a small force? We need to pull back all of our forces to get them out of harm's way and to secure our borders. We must prevent lawlessness and chaos at home."

"If you destroy the Islamic world with nuclear fire," Scofield said, "you won't be able to contain either the collateral damage that the radiation will cause or the likely retaliation of other nations. Such measures must be used only as the last resort, not as the first measure of retaliation."

Selected to meet with Scofield because he was a former Marine and might be able to trade on their comradeship, the governor was nonplussed by the general's lack of subtlety. *Okay, bluntness it will be, then.*

The two men were in the general's private stateroom. Both knew why the governor was there. He'd been sent by the hardliner faction to determine where the legendary general's sentiments lay: with them, or with the "appeasers."

"Maybe, General, but my people paint a different picture. If we don't strike a fatal blow, then America will die the death of a thousand cuts. We can no longer be idle or passive. We must make war and engage the whole of the Islamist world in that war." The governor was focused, intent on converting Scofield, and a true believer.

If the general hadn't known better, he might have agreed; after all, war was his profession. What the general knew, and the governor didn't, was that there was a larger purpose to Scofield's being there, to the attacks on America and their aftermath. "Is that what you mean by 'out of harm's way,' Governor?"

"We cannot attack while you are traipsing around the Middle East on your suicide mission."

"Well, if you believe that, why not just wait until we are dead and then attack? I can be your Gen. Custer, and our dying field a new Little Big Horn."

"We've considered that."

The general smiled. Leave it to a jarhead to find the sick humor in any situation. Though pitted against each other, the two Marines would always share a bond, and dark humor was a part of it.

Both loved their country; both believed they were right; both were willing to die for their beliefs.

"Are you aware of what's happening at home with the selection of the new president?" the governor asked.

"I get daily briefings," replied the general. "In situations like this, I recommend recalling your original oath, the one you took when you were commissioned a second lieutenant, the one you took when you assumed your office: 'To support and defend the Constitution against all enemies, foreign *and* domestic.'"

"I know my duty, General; do *you*?"

"Yes, Governor, I do, on every level."

"What are you going to do, General?"

"Attack the heart of my enemy, draw him out into the open, and kill him— every one of them, without mercy or remorse."

"How are you going to accomplish that with only ten thousand men, General? Are you mad?"

"Perhaps."

"You took off on this escapade without national authority, without the backing of the American people. We can pull your Naval support at any time. You work for *us*, damn it, and don't forget it." The governor had his answer. He knew that Mike Scofield wasn't behind them and would not obey their orders. There was little to do but try to slow him down, try to derail his mission until they could solidify power at home, and replace him.

A lingering fear of the governor and his compatriots were the persistent rumors that Gen. Scofield had new super weapons that only he controlled. If so, that could make Mike Scofield a dangerous man who might even have to be eliminated. On the other hand, maybe these new weapons could provide the advantage Jones' group needed.

The general stood up and motioned the governor toward the door. "Well, we understand each other then. However, before you and your group make any decisions about our fate, I would beg your patience a little while longer. Please stay another day. I have a demonstration of our technology planned for you and Sen. Fitzpatrick. Can I count on you, Governor?"

Sen. Fitzpatrick was a leader of the progressive faction. *So, the general has chosen his side,* thought the governor; but he would not miss the demonstration for any reason. Finding out what magic the general possessed was the second-most important reason for being there.

"You can count on me, General," he answered. They shook hands and parted.

THE FLIP SIDE

"Sen. Fitzpatrick, how are you today?" Mark asked. "You don't know me, but my name is Mark Osborne. I write for *U.S. News and World Report.*"

"I didn't know you from Adam until I saw your article breaking the news about where the general was going to invade. You caused a great brouhaha with that article, young man. Do you think it was prudent to disclose top-secret plans to the world while we are at war?"

"Just doing my job. I think the general wouldn't have told me if he hadn't wanted me to report it."

That made sense, but the senator, frowning, refused to acknowledge it. He suspected that Scofield had leaked his strategy to commit America to his purpose, knowing that the president-to-come would not be able to back away from it. *Smart bastard,* Fitzpatrick thought. *Damn him!*

"What are you doing here in the ward room, Senator?" Mark asked.

"I can't say. Why are *you* here?"

"Now *that* is a very good question, Senator. I can honestly say I've no idea." Osborne was still puzzled about that. Was the general playing him for a patsy? He hoped not. "Are you going to the technology demonstration, Senator?"

"That's classified," the senator replied, moving to another table.

Mark was about to follow when Gen. Scofield walked in and greeted the senator warmly. Unable to hear what they were saying, Mark watched as the two men strolled out of the ward room together, learning only later that the general had invited the senator to dine with him in his stateroom.

THE APPEASERS

"Thank you for rescuing me from that twit of a reporter, General," Sen. Fitzpatrick said as they reached the general's stateroom.

"He is harmless enough, I think."

"How could you let him report on your strategy? How is it possible that I find out this foolishness from the press and not from you? How can you fail to secure approval for such a strategy before announcing it to the world? That was just plain stupid, General. Are you running rogue? Can we, can your country, trust you?"

"I understand your anger, but it was necessary to commit to a strategy. Besides, my strategy requires that we lure the fanatics out from their hiding places, draw them into our trap. If we hope to do that, we must give them time to make ready their defense of the holy cities, time to travel from their cockroach-infested lairs to defend their faith. This will require that we provide the Templar Division as bait. We must be prepared to risk everything."

"You don't have the authority to commit the United States to any strategy! We are a constitutional republic, and the military follows civilian authority!"

"Which civilian authority do you wish me to follow? Last I checked, they were all dead."

"The Congress has the authority to appoint the next president; we are in the process of doing that now. You should've waited until that process was complete before you struck off on your own and announced plans to invade the world."

"Seven senators and a handful of congressman cannot reasonably appoint the next president, Senator. The governors want you to wait until they've appointed replacement senators and congressman so that all fifty states are represented. Anything you do unilaterally will be seen as illegitimate."

"If we wait for those filthy Nazis to appoint their parrots, we'll lose our democracy to right-wing fanatics. Besides, they're taking forever to make their appointments, and we need a new government now."

"We both serve a representative republic, Senator, even if you don't like the men and women across the aisle. If you overstep your authority, you may lead us to ruin, even to civil war. Both sides need to compromise."

"Is that what you told Governor Jones?"

"No. I told him to go screw himself."

Mulling that over, the senator responded, "So, what are you saying, General?"

"Allow the system to work, don't undermine it, and let me do my job."

"I can't make that commitment right now."

"I understand, Senator. Meanwhile, please stay a while longer. I have a demonstration of our new technology I'd like you to see. It might allay your fears about the folly of my enterprise."

A NEW WORLD ORDER

Mark entered a cargo hold on the ship with his female Marine escort. She was nice enough, even comely—for a Marine in an unflattering uniform. The baggy cammies could not conceal her curves. She told Mark that he was the only member of the press allowed at the demonstration and reminded him that it was highly classified: If he reported anything without specific permission, she would eviscerate him. She wasn't flirting.

An area had been cleared in the center of the hold, and a glass cage erected there. Mark and some military personnel were ushered to folding chairs on cheap, raised platforms. From a seat in the back, Mark spotted Sen. Fitzpatrick and Governor Jones in the front row. Behind them sat the military brass, most of whom Mark didn't recognize. Gen. Scofield was nowhere to be seen. The whole thing looked inexpensive and thrown together; the only high-tech element was an array of large plasma displays set to one side.

A figure entered the glass cage, he wore the same form-fitting garb reporters at Camp Lejeune had seen on the parade grounds, and was obviously muscular and fit. His entire body was encased in a smooth black exoskeleton, one Mark had never seen. Seamlessly attached to his "battle suit," his helmet completely covered his head, with a multi-angled glass-like face screen. It looked as though his suit had been poured onto him like molten plastic. He was simply outfitted: a belt with a small pouch, a sword slung across his back in a sheath, and a set of strange forearm contraptions that Mark could not make sense of.

This warrior carried none of the usual combat "kit" that weighed down the average modern soldier.

The soldier's opaque face shield turned clear—it was Gen. Scofield. The plasma screens came on, displaying images sent directly from the general's helmet, showing the onlookers what the general himself was seeing.

Mark was fascinated. The main screen displayed a composite picture of visuals from many different sources, blending ordinary video with thermal images of the spectators and night-vision illumination of the dark areas behind them. Computer icons—some familiar, some not—were superimposed on the screen. An x-ray feature captured the almost perfectly nude image of the female Marine who had escorted Mark into the cargo bay, displaying her on a separate flat screen as well as a picture-in-picture image in the general's main display. A selective feature, apparently, since no others were so exposed.

The Marine's beautiful body was perfectly formed, shamelessly unrobed, disturbingly erotic. Her nakedness clearly didn't faze her, but her fellow spectators took pains to appear nonchalant. Scofield let the crowd linger on her image, then turned off the probe and began to speak: "When I assumed the office of Assistant Secretary of Defense for Science and Technology, I started a program to evaluate the effectiveness of our weapons systems' programs on the battlefield. I decided to look at the programs from a simple evaluation perspective. For each dollar spent, what was the return, in real dollars, in battlefield effectiveness?

"As expected, programs like smart bombs performed well under such analysis, since they allow us to vastly increase our destructiveness while decreasing our cost. We also factored in the cost of indiscriminate bombing when civilians were inadvertently killed or maimed. This latter cost analysis showed a significantly increased savings for smart bombs, as our country and the world had become less tolerant of such killing. This added benefit allowed the United States to use this form of bombing in a broader set of circumstances. The benefits, in short, were almost incalculable.

"Other programs, like new bomber designs and new missile systems, didn't fare as well. We found that we were spending vast sums of money on systems that we rarely used. Not surprisingly, at least to me, was the cost analysis showing that boots on the ground, individual combat soldiers, were the largest cost center in modern warfare. Not because the ground troops' combat kits were particularly expensive, or even because there were so many of them—although the cost of supporting ground forces in beans, bullets, and band-aids *was* very high. The highest cost appeared in a detailed analysis of soldiers and Marines lost to wounds, death, and dismemberment. This was particularly true in Iraq during the peak of the battle. The high cost of combat injuries was multifold: direct costs associated with evacuation and treatment, payments to spouses, rehabilitation, and long-term care. But these were negligible compared to the costs of required tactical changes, overbuilding combat capability to assure superior force during an insurgency. Having boots on the ground inevitably required more boots, and casualties required even more. We were spending billions every month fueling the fight against the insurgency.

"We clearly needed a smart-bomb equivalent to make the fighting soldier more effective. As with smart bombs, a significant breakthrough would allow us to use ground troops more freely and with less risk. So, we started a program to evaluate current technology to see where we could make a difference. Both the Marine Corps and the Army had substantial programs to develop better combat kits for near-term warfare needs, but they seemed to miss the point. They focused on short-term point solutions to provide better targeting, better communications, or better survival. These systems were not revolutionary, and not well integrated. Plus, their combined physical weight, if deployed together, would have been staggering. We initiated programs to evaluate technology with breakthrough potential—robotics, mechanized power suits and nanotechnology, to name a few. We decided quickly to focus on nanotechnology, the science of miniaturization. Integrated systems only nanometers in size –that's where we would find our breakthrough.

"However, when we evaluated the state of nanotechnology research, we were not pleased. Nonetheless, we felt that an adequately funded program focused on our needs could lead to paradigmatic shifts in infantry combat. The cost of such a program? No more than a new bomber design or aircraft carrier—a good investment—and so we began.

"Our preliminary studies showed great promise, so we embarked on a top-secret program to develop nanotechnology-based warrior armor, weapons, and integrated communications. The program was kept in deep-compartmented security, more secret than anything since the Manhattan Project that developed the atomic bomb and ended World War II.

"I'm wearing the results of that research. This armored fighting system makes our soldiers and marines nearly invincible on the battlefield. We can now fight our enemies with far fewer troops, suffer few casualties, defeat any foe, and project power effectively in ways that would have been unimaginable in earlier times. The World Order has shifted."

He paused, letting his message sink in. "You have seen the effects of our invisibility—or, as we like to call it, our cloaking layer. We have many other layers, equally magical. Our armored layer is impervious to small caliber blunt trauma, from bullets or shell fragments. It deflects the kinetic energy from the projectile around the suit and expels it to the rear. Much of the impact energy is collected, converted, and stored in multiple battery layers in the suit. The more our troops are hit, the more energy they have to fight.

"The suit has multiple layers to address environmental concerns: one to process feces, another for urine, another to warm, and another to cool. One layer fits directly against the skin, acting like a gentle exfoliate and moisturizer. The suit leaves your skin soft, cleansed, and refreshed.

"Another layer, with male and female versions, now discontinued, would auto-pleasure the soldier, but it led to unsustainable losses as warriors wandered off dazed into the woods."

The crowd laughed at this apparent joke, but Scofield assured them that it was mostly true. The pleasure system had been too effective, and warriors had become addicted to it. Scofield told them that the suit was inherently constricting, especially when worn for days in the field; so some method of stress relief had been deemed necessary.

Scofield continued: "The suit has a layer to block microorganisms and chemical agents that reacts automatically whenever a contaminant is detected. It has a layer that projects sound energy—you heard some of that the day we embarked. The sound can be a beautiful melody, or a crushing weapon when combined with many suits and harmonized to a killing frequency."

"I know you made this ship invisible, are there plans to expand the coverage and add things like radar suppresion to the Navy ships of war?" One of the admirals in attendance asked.

"Yes, absolutely, we are working on that as we speak." The general responded, then he continued with his brief.

"There are multiple layers that capture the motion of the warrior and convert it into stored energy for later use. A warrior's heartbeat can power the suit. Similarly, external energy sources are captured, like wind and sun, to provide additional energy.

"Another layer aids the neuromuscular system, providing the warrior with enhanced strength, speed, and reaction time. This was one of the most difficult layers to develop, since it tended to exceed the strength of the human frame, snapping bones like twigs. We dialed back the system, using biofeedback to let the layer know when too much is too much.

"The cloaking layer can do more than hide you from your enemy. It can also project images, because each nano-element in this layer is the equivalent of a tiny pixilated transmitter, a tiny multifaceted television set.

"Perhaps the most important component is the nanolattice fabric layer, consisting of independent nanorobotics. These nanobots are typically much larger than the nanobots used in other layers for specific effects like cloaking. They're bound together in a three-dimensional fabric that manages all the distinct nanolayers. The fabric may be connected in a flat two-dimensional manner, or extended to many layers as needed. Each lattice nanobot has tendrils allowing it to be bonded with others in any three-dimensional matrix. Each has an X, Y, Z spatial assignment, and a gigabyte of memory storage. Each nanobot has a serial communications highway that runs at one terabit per second along its tendrils, and an internal processing engine housing an enhanced Turing-Machine-based computational engine."

"You're losing me General, what's that got to do with anything?" Sen. Fitzpatrick asked.

"Bear with me Senator and I'll try to piece it all together during the practical demonstration part of the brief." The general replied.

"The billions of nanobots in the lattice fabric provide a distributed computational engine that has greater cognitive power than any super-computer ever built. Throw a billion nanobots into a pot and they will self-organize into their exact programmed configuration within ten seconds.

"This intelligence and this spatial awareness allow us to control, precisely, who can wear the armor and who cannot. If a spy were to steal a nano-armored suit and put it on, the suit would kill him—with prejudice."

"Who controls the programming of these suits, General?" One of the admirals in the audience asked.

"I do."

"Is anyone else privy to the keys to this magic?" This time the question came from Governor Jones.

"No.

"I'm now going to demonstrate the efficacy of this armored suit."

Warriors in nano-armored suits placed three men, dressed as Afghan mujahedeen, in the glass cage.

"These men are captured mujahedeen fighters. They cannot see or hear you, and they cannot hear me, although I've allowed them to see me. To them, the glass cage is a solid walled room, the effects of its cloaking layer reversed. You can see through it, they cannot."

The mujahedeen looked wary, menacing. Three AK-47s, two grenades, and numerous swords and knives appeared when Scofield removed a cloaking blanket from a table between himself and the fighters. The mujahedeen didn't move.

The general spoke in Pashto, his words translated on the display screens in streaming text:

…MY NAME IS GENERAL MICHAEL SCOFIELD. YOU KNOW ME AS THE LION. I AM YOUR SWORN ENEMY. I INTEND TO KILL YOU TODAY, UNLESS YOU KILL ME FIRST. YOU HAVE NO CHANCE OF SUCCESS. I AM PROTECTED BY THE SUIT I AM WEARING. STILL, TO DIE FIGHTING IS BETTER THAN TO DIE A COWARD. I WILL GIVE

YOU ONE MINUTE TO GRAB YOUR WEAPONS AND FIRE. AFTER THAT, I WILL KILL YOU WITHOUT MERCY. MAY THE LORD GOD HAVE MERCY ON YOUR MISERABLE SOULS...

The mujahedeen looked at one another, praised Allah, and grabbed weapons. The ruckus that ensued was spectacular. Two of the fighters picked up AK-47s and sprayed the general with lead. The third fighter grabbed a grenade, pulled the pin, and threw it at Scofield. The general's suit repelled the bullets like a force field, as did the walls of the glass cage, with no apparent ricochet. The general caught the grenade in his left hand like a baseball, smothering it as it exploded, thereby saving the lives of the mujahedeen. In awe, the three stepped back, waiting.

The Lion of Afghanistan, Gen. Michael Scofield, pointed at one of the three. With a soft hiss, a bullet shot forth from his forearm, exploding on impact in a violent flash. The effect on the targeted mujahedeen was horrific. The mujahedeen was cut in two—his midsection vaporized, his upper torso thrust into the ceiling, his lower body crushed to the floor.

As the onlookers were registering the shock of the first bullet's impact, the general shot the second mujahedeen fighter, killing him in similar fashion. Then he stopped, staring at the third and final victim, who was frozen with fear. The general removed his sword and waited.

Realizing what the Lion wanted, the mujahedeen, through sheer toughness, moved to the table and grabbed the other sword. With bravery few of the spectators could have mustered, he charged the Lion, sword raised, screaming his Pashto war cry.

The general's sword glowed eerily as he swung it toward his attacker. He sliced through the enemy's sword like butter, through the mujahedeen's shoulder, through his torso, and out through his hip. The resulting mess was ghastly. He lay in two parts, his innards pouring out in a gusher of blood.

The general sheathed his sword and walked through the glass wall.

"As you have seen, we've perfected our killing weapons as well. We no longer carry rifles that shoot kinetic energy bullets. We've replaced kinetic weapons with tiny chemical bullets that mix into an explosive brew when they are fired and hone in on the target we've sighted. Nanofins guide the round to its destination. Point the round in the general direction of the target; it will go where you were looking. Explosive parts are not mixed until it's shot; the round remains inert and harmless while stored. Each dart is so small; we can carry tens of thousands of rounds. And, given the automatic targeting, we can achieve a near one-hundred-percent hit rate. One warrior can kill a thousand men, with little risk to himself and without resupply.

"The sword has a nanolattice core, with diamond-sharp cutting nanobots running along its edge at one thousand rotations per second like a miniature chain saw. We have other tricks, but that's enough for now. Any questions?"

The general was met by stunned silence. The spectators had just witnessed a four-star United States Marine Corps general murdering three helpless souls. He hadn't merely murdered them; they'd been slaughtered, mercilessly, just to make an impression.

ON THE LEFT FRONT

"What do you think?" Sen. Fitzpatrick asked Roger Matrick.

"I honestly don't know. If he has half the technology that he says he has, then everything has changed, and he is the most powerful—and dangerous—man in the world." Roger shot back.

"Yeah, maybe, but how will he use that power?"

Roger Matrick was a major political operative in the Democratic Party and a major player in the loosely defined "progressive" group pushing for quick action to reconstitute the government. Fitzpatrick was reporting to him on his visit with Gen. Scofield.

"The way he murdered those men in front of all those witnesses tells me he doesn't care what we think," Roger observed. "It's scary that he would do that, his gestures always have a purpose. He may be mad, but he's as smart as anyone I know."

"His purpose was obvious: he wanted to scare the hell out of us—and I, for one, can tell you he succeeded. I can't get those poor bastards out of my mind." Fitzpatrick said.

"War is hell. He has done much worse on the battlefield. I can show you the videos to refresh your memory." Rogers replied.

"On the battlefield, he was risking *his* life, too. He had no risk in that demonstration. He didn't have to kill them to show how ineffective their weapons were. He could have demonstrated the power of his own weapons another way—destroying an inanimate object, for instance, anything but what he did. He may be mad. If so, he's a madman with multiple weapons of mass destruction!"

"Has he aligned with the hardliners?" Roger asked warily.

"My intel says no. Why should he align with any side? He can just declare himself President! What the hell could any of us do?"

"Perhaps, but let's not panic. Maybe he's not mad. Maybe he's as divine as some say he is, and when he's done punishing the Arabs, he'll come home and make you President." Roger tried to calm Fitzpatrick as he furiously calculated how to exploit the new political landscape. "In the meantime, we need to throw our support fully behind Scofield. That would be our best strategy at this point."

"Agreed."

ON THE RIGHT

"Sounds like you had a rough trip." Governor Blake Tilden was debriefing Governor Ray Jones on his visit with Scofield.

"He let me make an idiot of myself," Jones responded. "I told him he was a fool to go to war with ten thousand men. Now I know he could defeat any army on the planet with a few hundred wearing those suits."

"It sounds like he's prepared to put on quite a show. I personally can't see the downside, though. If the United States has the most powerful army in the world, then let's unleash it, and we can achieve our same goals without the messy radiation and the complaints from our neighbors."

"Yeah, then what? He comes back here and disarms? He has the only keys to this technology. Why would he give up that much power willingly?"

"Because he believes he's the anointed of God," Blake Tilden said with a smile.

"Oh, yeah, I forgot!"

"Meanwhile, let's find out more about this technology. Who do you know in the Department of Defense? Also, signal to Scofield that he has the full support of the American people and our group in particular."

MAKING SENSE OF WAR AND DEATH

"Good morning, General," Mark Osborne said, halfheartedly. He was no longer in awe of the general—just deathly afraid of him.

"Is there anything you want to talk about, Mark?"

"Why did you have to kill them—like that?"

"They were evil men who deserved to die, for many crimes: rape, murder, extortion, terrorism. The list of their sins was endless."

"So you were their judge and jury?"

"Yes. But if it would make you feel better, they were already sentenced to death by a military tribunal."

"We don't execute convicted murderers by cutting them in half."

"Not typically, no."

"May I be excused?"

"Of course."

THE ANGEL MARIA

"We had a hard time finding her because she had already left the convent for the ship," a member of Gen. Scofield's staff said. Scofield and his aide were discussing one of Scofield's former members of his Special Forces unit in Afghanistan and Iran, Maria Olsen. "When she heard you were going to war, she immediately packed her trash and came after you. My guys tell me she should arrive any time now. They put her on a fast mover to Italy. She will transfer to the ship via a V-22 Osprey as soon as we get close enough."

"Thank you, Major. That's welcome news."

Maria was more than his former aide. They had first met at an Afghan military hospital near Kandahar, where he was recuperating from yet another Purple Heart. She was a Marine sergeant, sweet and innocent, assigned to the hospital as a supply clerk. While he was there, the base and hospital were attacked and overrun by Taliban fighters. Maria and then-Col. Scofield were thrust together in a desperate fight to protect the hospital staff, a battle that became legendary. In its aftermath, Col. Scofield would be awarded his second Medal of Honor. (To make this possible, Congress would change the 1917 law limiting awards of the medal to a maximum of one.) He would also collect Purple Hearts fifteen through eighteen in that two-day action.

Maria was a young girl, just twenty-two years old, and not long removed from the Kansas trailer park outside Topeka where she had grown up. Caught up in the whirlwind of combat at the hospital, she was pressed into service reloading for the few Marines fighting in the trench just in front of her. Maria found herself supporting four young Marines and one battle-scarred, mean-as-hell Marine Corps hero nicknamed the "Lion." She could never have guessed where that association would lead. From that date forward, whenever he was in or near combat, Maria would not leave the Lion's side.

The general went to sleep thinking how nice it would be to see Maria again, wondering how she had changed over the past two years. He worried that she might not be up to this new challenge since making her spiritual retreat into a cloistered convent.

The knock at the door was so soft that Mike barely heard it. It wasn't a wake-up call—those were noisy. It must be Maria.

"How was your trip?"

"Long. Go back to sleep. I'll grab a couch, and we can catch up tomorrow."

"Come to me. Let me see you."

The Lion's eyes had been enhanced; he could see well in darkness. They simply took each other in, smiling softly, welcoming each other home. Maria was forty-five now but could have passed for a teenager. Her skin was flawless.

"You look well, Maria."

"That's easy with the alchemy you give me. I see that you have been drinking the same elixir, my dear."

"Was it hard leaving the convent?"

"Yes, and sad. The nuns were heartbroken. It was doubly hard for them, knowing where I was headed. They honestly cannot fathom how anyone can do what we do."

"That's understandable. I was surprised to hear that you had entered a cloistered nunnery. I guess it makes sense, given your journey. Still, it made me sad. At least you had the haircut for it."

Maria had shaved her head while serving with Scofield to fit in with the other fighters in the Lion's outfit. She'd been the only woman in the group at the time. Not long after joining the unit, they'd become lovers. She'd been a lost soul after that desperate fight at Kandahar, clinging to Scofield like a lifeline. She would do anything for him. After she'd shaved her head, they both decided that they liked the look. Her smooth, perfectly shaped head was attractive on her and was pragmatic for a long combat tour in Afghanistan.

"You were always a passionate woman. How did you endure a life of abstinence?"

"After Iran...I needed to heal."

Mike nodded.

"Asceticism became part of my journey."

"Did you take your vows?"

"I arrived with no expectation that they would accept me. My reputation as a warrior, and from their perspective, a sinner, preceded my arrival. I sat down with the Mother Superior and we talked for the better part of a day. She is the kindest person I've ever known, insightful and empathic, like you when you're not being cruel."

Scofield nodded.

"She decided to let me stay with the sisters, not as a novice, but as a guest. As far as I know, that was the first time they had let anyone do that. I think she saw me more as a patient then a potential recruit. She was a psychiatrist by training, so maybe I was her project."

"Did you tell her about your experiences in combat?" Scofield queried.

"She was an expert at asking just the right questions. I was very emotional at the time, near despair. I opened up to her and recounted my experiences fighting with you and the boys, our lives of danger, killing, death, and passion."

"She must have been shocked to hear your stories, I've never told anyone, not even my family—what happened to us—the things we did," Scofield admitted.

Maria said, "She was a good listener. But for me, speaking the words, reliving those memories, was very purgative. She became my confessor. There were times when tears would flow down her face—just a gentle stream of tears as I recounted horror after horror. I felt forgiveness, mercy and acceptance in her presence."

"Did you live as a nun during your time with them? Accepting celibacy, denial, and poverty, without saying the vows?" Scofield asked.

"Yes, willingly. They taught me that celibacy, when coupled with a contemplative life, was transformative. We raised our voices together in song many times a day, and our souls were lifted as one. We worked hard to perfect our harmony. I also learned that a small kindness was an expression of love. I put my passion and rage to sleep for a time."

"And?" Scofield probed.

"And…our enemies returned, and with them, my rage," Maria said.

"Just like that? Snap, and you are a no longer dedicated to a life of peace and reflection?" Scofield probed.

"I had dreams, months before the explosions—dark dreams—that foreshadowed my presence here. There was an awakening, I understand that now." Maria's voice hardened. "I believe these dreams were God recommitting me to my duty as His warrior. He gave me time to heal, then handed me sword and shield and sent me here."

"Are you ready to do what's necessary? To do your duty standing with us?" Scofield asked.

"Yes."

"I'll make you a colonel then, and put you on my staff." The general made a mental note to update his chief of staff.

"What about us? Have you committed yourself to a life without sin?" Scofield probed, both knowing his meaning.

"To deny you would be the same as denying the air I breathe. I've never had the strength for it." Maria looked down, feeling helpless in his presence.

"My heart isn't pure like yours, Maria. I know our infidelity is a deep wound for you, and I am ashamed that I don't feel the same guilt. My wife has forgiven us both, but I know it wounds her. She lost her husband and partner when my soul was blackened by war and then later my fidelity. I don't deserve her or you."

"I understand Sarah better than you do because you cannot see the power you wield over women. You are right though—we don't deserve her. Sarah's forgiveness of my betrayal is a scarlet letter I must bear…her love, a knife through my heart. Why do you think I left you?" Maria looked sad, remembering her crisis two years before.

"We are lost to our sin then. God forgive me, but I do love you so," Scofield's arousal was obvious.

The conversation roused old feelings within Maria. She hadn't seen Scofield for years and longed for him.

Maria silently disrobed. She was stunningly beautiful, her body exquisitely muscled, long, and lithe. She was perfectly alluring, and an expert at giving pleasure. Mike loved her and ached with anticipation.

"Just hold me, Scofield. There will be time later. Just hold me…" Maria quietly wept. She was lost again to this world of war and death and longing for a man she could never fully possess. He'd never understood—her agony was watching him die a little each day.

GETTING READY

Preparing any invasion is a laborious task. Most of the preparation is typically wasted as soon as the first shot is fired, since all plans go straight to hell when the operation starts. Nonetheless, the process of preparing is cathartic, forcing planners and executers to think through all of the details. Having been through this process, they can better react to changes when they are required. Still, to most untrained observers, even a well-planned invasion appears chaotic.

This invasion was different. The technological advantages were so overwhelming that the planners were understandably lax in working out the details. Scofield's simple plan was to land at the port at Jeddah and march his forces under fire through the town, then east on the Alharamain Expressway to the outskirts of Mecca. The general would then steer his division to the Jeddah-Makkah Expressway and into the heart of the holy city. After occupying the holy Mosque at Mecca and provoking the whole Islamic world, he would assume the high ground near town and await the angry armies and hordes that would soon seek his death.

Scofield had commanded his logistics managers to plan pre-positioning, resupply, and support services. This part was tricky. The port at Jeddah, so deep into the narrow Red Sea, was hard to defend. The Navy would be hard pressed to secure the beachhead and might have to withdraw. The general was prepared to be cut off and continue fighting. He'd had the foresight to work out the energy-collection capabilities of the nano-armored suits. They could live off the sun, drink their processed urine, and eat their processed feces if necessary. *Hell, we'll barbecue the dead if it comes to that,* he chuckled to himself. But it would not come to that.

The armada carrying his nano-armored Templar division would follow the western contour of Africa and would eventually round the Cape of Good Hope at its southern tip. Before reaching the Cape, the small fleet would be joined by a carrier battle group that had left Norfolk just ahead of the amphibious force. With that addition, the fleet would more than double its size. Two more carrier battle groups awaited them on the other side of Cape Agulhas, just east of the Cape of Good Hope at the southernmost tip of Africa. The final fleet configuration would be a massive, overwhelming force unchallengeable by any two navies around the world.

To hedge his bets, Scofield had sent nano-engineers to the battle groups to "paint" the ships' exteriors with lattice work for cloaking and armor. The fleet would be invisible and, hopefully, invincible—although the armoring of ships was largely untested.

The senior staff was assembled in the command center on the general's Flag Ship. Adm. Michael "Mad Mike" Franklin, Scofield's four-star naval counterpart in the operation, sat directly opposite the general. The two men shared a lot of history. Mike Franklin was a former Navy Seal who had participated in combat operations with the Lion in Afghanistan and had later fought alongside him in Iran. Also a Medal of Honor recipient, Mike Franklin had a great respect and affinity for Scofield. It was unusual for a Navy Seal to reach the rank of admiral—especially a four-star fleet admiral like Mad Mike. He was there because he was the best there was, and because Mike Scofield wanted it that way.

Seated to the general's left was Maria, now in the uniform of a Marine colonel, and his senior staff: Colonels Tad Johnson, Bob Thornton, and Miles Franks, the commanders of his three operational regiments. To his right was his chief of staff, Col. Melissa Franks, Miles's wife; and the general's executive officer, Brig. Gen. Tommy Marks. The general's staff consisted of battle-hardened men and women, the very best the Marine Corps could muster. The warriors at the table had served with Maria in combat and knew she had become a serious and deadly warrior, a person worthy of respect who warranted a position at their table.

The admiral was accompanied by one Navy captain, his chief of staff, and nine admirals of various ranks. The naval fleet they were assembling was vast and required significant rank structure, with the naval personnel outnumbering the Marines three to one.

Adm. Franklin spoke first. "General, we've looked at your battle plan, and we would like to suggest some changes. We're very concerned that we won't be able to defend the beach at Jeddah once your troops have disembarked. We'll be exposed to land-based air attack from almost three hundred and sixty degrees, and the fleet could become isolated. We could be cut off if the enemy deploys enough naval power against us at the entrance to the Red Sea. We can expect the Suez Canal to be closed to us."

"Perhaps it would be better if we attacked at Aden in Yemen and established a sustainable base of operations there, in order to provide better support for you and maximize our ability to protect the fleet. I know that this would vastly increase your operational extent, and that your supply lines would be stretched and vulnerable, but with our technological advantages, this might not be a big problem."

Gen. Scofield responded. "Admiral, we expected you to request this. When I pondered the difficulties of this invasion, I knew the Navy would request this approach. Hell, it makes sense, I agree, but the answer is no."

Though the admiral's face reddened, he had expected this from Scofield. "General, this is just the first battle of a world war. You are going to need my Navy to fight this war, and what good will it be at the bottom of the damn Red Sea? If all your Navy buddies are dead, how will you get home, General?"

"Feel strongly about this, do ya, Mike?"

"You betcha, Lion. And I'm not letting go easily. If you want to risk thirty thousand Navy lives and over two trillion in hardware, then I want to know why."

"We must use this force to provoke our enemy. I intend to insult the sentiments of these Islamic bastards and give them no chance to avoid my trap. We don't have the resources to traipse around Arabia, Persia, and Africa to find these rats. Our choice is to kill everyone in a nuclear holocaust—which a strong faction in our government wants to do—or be more selective in who we slaughter. Which of these two paths would you rather follow?"

"Your path, of course," said the admiral, "but I don't see why that path excludes my approach."

"Fair enough," the Lion replied. "The truth is, my rationale requires a certain leap of faith. By steaming into the Red Sea, guns blazing, we are challenging the very manhood of Arab nationalism and Islamic pride. We're saying we no longer give a shit about your Islamic sensibilities, and we are going to punch you straight in the mouth. If you want to fight us, fine. Come out and fight, because we are launching the tenth crusade, you sons of bitches, and you can't stop us. I named this division the Templar Division to evoke a crusader image—and, frankly, to piss them off.

"We'll station the entire fleet near Jeddah, except for enough ships to guard the entrance to the Red Sea, and we'll defend our beachhead to the death. It won't come to that, given the armor we are applying to the fleet, but the enemy doesn't know that. They will break their air and naval backs on our ships while we slaughter their armies on the ground.

"The effect of our naval presence in the Red Sea and our division sitting on their holy sites will be to galvanize our enemy to action. In that action will be their doom."

"It will only take a handful of nuclear weapons to sink the fleet in those tight spaces," the admiral countered. "Pakistan has enough by itself to destroy us. Can your armor save us from a nuclear explosion?" The admiral was beginning to wonder if maybe the old guy was crazy.

"We haven't tested the armor at those temperatures and blast waves. Theoretically, yes, we could survive. Truthfully, we just don't know. As for Pakistan, we have a plan to neutralize their nukes. That's in the works as we speak."

"I cannot change your mind, can I?"

"No, old friend, you cannot," the Lion replied, "but I'll avenge your death if I'm wrong. You have my word on that." The general winked at Mad Mike Franklin. "Admiral, wear your medal into combat, as will I."

"May God have mercy on our souls, General."

"I guarantee it, Admiral," the Lion asserted, but his professed connection to the Almighty left Franklin uneasy.

"One more thing, General. We've been tracking four submarines shadowing our fleet: two from China, and one each from India and Russia. What would you like us to do with them?"

"Sink them, quietly and simultaneously. That should give them pause, should they decide to threaten us as we get into the shallow waters of the Red Sea."

"Yes, sir. Should we deploy the sea jets?

The general nodded. "Time to get them bloodied, I think."

MARK'S TAILSPIN

Mark had to admit that he was cracking up. He had seen too much, too fast. His visions troubled him, of the girl—seeing in her eyes the reflections of millions of dead Americans back home—and those poor Afghan men slaughtered by the general. He was haunted by two images of opposing evil.

Mark had not ventured out of his stateroom for the last twenty-four hours. He was too depressed to function. But his self-pity had to end. He finally succumbed to the call of hot chow. It was time for mid-rats, the hot snack put out for the crew around midnight.

Mark entered the room, grabbed a beckoning ham sandwich, picked out a table where he could be alone and sat down to eat.

"Hello, Mark. My name is—"

"Maria Olsen. Yes, I know."

"May I join you, Mark?"

"Of course. I'd be delighted."

"Mike has told me about you. He's taken an interest in you," she said sweetly.

It felt strange to hear her speak so informally about the top dog on the ship. "Do you mean he takes an interest in lost souls?"

"Something like that. Mike has a keen insight into people. It's as if he can see into a person's soul and judge whether they are good or evil."

"What did he see in my 'soul'?"

"He thinks you are a good person—just a little lost," Maria said.

Mark marveled at Maria's soothing, disarming personality. It was impossible not to be drawn toward her, to want to be taken into her confidence. *Being beautiful doesn't hurt either,* he thought.

"Is that why you came over here? To pet the little puppy and tell him it's going to be okay?"

She smiled. "Scofield sensed that the demonstration disturbed you. He—"

"The *demonstration*? You call that a 'demonstration'?"

"What would you call it, Mark?"

"Murder. Cold, calculated murder."

"At some level, all killing is murder. War is legalized murder if your side wins and genocide if you lose...and survive to talk about it. Perhaps it's a convenient ambiguity."

"Some worship you as a kind of angel, but you've killed men in war."

"Many poor souls have died at my hands—in some cases, my *bare* hands."

Mark paused at this remark. He thought, *a beautiful woman, sure, but also muscular, lithe and, he struggled to put a word on it—cat-like, predatory in a weirdly erotic way.* Mark shivered involuntarily. "How do you reconcile that with your 'spirituality'?"

Maria paused, contemplating her response; "God has sent me on a mission. I've often questioned why He would ask me to do these terrible things and cause me so much pain. I don't have the insight to tell you why He does what He does. I'm a simple messenger. I cannot reconcile my actions in war with my faith in a loving God and the message of peace from his Son. In the end, though, there is only faith or despair—and I choose faith."

"I haven't been able to get those men's deaths out of my mind. Why did he kill them like that?"

"I wouldn't ask him, and he didn't elaborate."

"You have unquestioned faith in Scofield, then?"

"Yes, I do. I love him and trust him. He has saved my life so many times that I can no longer measure my debt to him. Even without that debt, I know that he has been touched by Him and has suffered greatly for the honor of it. Because of that debt and homage, I'm his vassal to do with as he pleases." Tears welled up in her eyes.

"He must have reasons for his actions then," Mark said softly.

"He has been touched by Him, but is a flawed man nonetheless. God chooses a man for reasons of His own, but leaves him free to make mistakes—even terrible, painful ones. Mike Scofield will be judged by God like any other sinner. Maybe he was insane with anger," Maria suggested. "Maybe he was preparing himself for the slaughter to come. The masses he will slay in battle are likely to be less deserving of death than those three psychotic killers. They will be patriots: men and women of faith come to defend God's house. And yet, they will be slaughtered, too. Which is murder? Which is war?"

"How can you be a part of this?"

"I go where he goes. My life is his."

Maria invited Mark to continue their discussion in her stateroom. She was willing to tell him more about the technology Scofield had helped create, but not in a public forum.

"Do you mind if I get comfortable?" she asked.

Oldest line in the book, he thought. "No, of course—"

The question, he realized, had been rhetorical. She casually slipped out of her uniform and took off her bra and panties.

Her body was perfect. Her nakedness, which she didn't rush to hide, was accentuated by her smooth, hairless body and perfect skin. Mark couldn't believe she was in her forties.

She slipped on a robe, "You look uncomfortable, Mark."

"Well, we don't really know each other."

"There was a time when I would have taken great pleasure in seducing you, starting with a performance like that. Knowing that you are gay would have made the challenge even more tempting, and exciting. Now, well, I no longer care about being naked—in front of men, or women. I've been with men in intimate situations most of my life. There is no greater intimacy than combat. I guess I've lost all inhibition and modesty."

"Is it common knowledge that I'm gay?"

"As you might expect, we've done detailed background checks on everyone who enters this inner sanctum. Yes, it's common knowledge. Look, Mark, nobody cares anymore. Some of our manliest and most-capable warriors prefer men to women. The incredible intimacy in fighting units makes for a randy environment, especially now that so many women have joined men in the front lines. Life is short, especially in this business. Besides, most of the people here are modified in one way or another."

"Modified?"

Maria answered slowly. "Mike wants you to gain a deeper understanding of what this time in human history means to all of us. He wants you to be able to teach the rest of the world about the new technologies. You need to leave behind your previous life as a reporter and begin a new life—as a teacher, a mentor. It's no longer about scooping your competition or rising to the top in your profession. If it's money you want out of life, Scofield can write you a check for ten million. He has more billions then he can spend."

"Why do you need *me*? Hire a historian if you want one."

"Your access to the media is important. You'll be reporting on some of the most momentous events in human history, but we need to know that you can do so in a measured, intelligent way consistent with our timetable. Can you do that?"

"Yes," Mark said. "Now, tell me about these modifications."

GENETIC SOUP

"When Mike took over as head of Science and Technology at the Department of Defense, he started the nano-warfare project he described at the demonstration. What he didn't say," Maria confided to Mark, "was that he also started a program to investigate the human genome."

She continued, "Mike was always a smart and resourceful engineer, but I believe he was given certain additional gifts when he was touched by Him— insights into genome construction and nanospace. I've been told that Mike played active and direct roles in both projects, guiding the scientists and engineers toward a solution.

"In the genome investigation, he developed three teams of researchers and built firewalls between them. He had a team of health science PhDs, biologists and MDs doing the usual investigations into cell genome interaction, genome modification strategies, and genome functional mappings. He fed their data and data from other researchers to another team of information researchers and data mining experts that was firewalled from the first group.

"The third team was a skunk works group composed of two very different types of specialists: NSA and CIA encryption-code breakers and a collection of the world's top computer scientists who specialized in reverse code engineering. They used the information fed to them by the data miners to investigate the way the genome "code" interacts.

"Mike ran the skunk works directly. He was the only one who knew each of the other group's tasks, assignments, and discoveries. He made sure that each group's goals meshed with the overall mission of decoding the genome's program.

"The program?" Mark asked.

"Mike's great insight was that the genome, of any species, was a type of computation. He wasn't the first to assert this possibility, just the first to really delve into its potential meaning. He believed that each gene acted like a binary element in a computer program, except that the genome uses a quaternary system, with four basic elements instead of two.

"The binary system of digitized ones and zeros is the foundation of all computers. The genome has four basic stateful protein elements that are represented by the letters A, T, C, and G. These elements are combined into DNA molecules, which are the stateful equivalent of a computer "word." In a computer, a word typically represents some concrete piece of information like a letter of the alphabet, an integer, or a real number. These computer words are combined in various ways to create programs, databases, and information. Mike believed that the human genome was a similar computation made up of quaternary elements combined into words, and words combined into a program. If we understood the genomic program, he thought, then we would understand God's secret *recipe* for life."

"But why does it have to be a *program*?" Mark asked. "Maybe it's just a random sequence of information strung together by billions of years of Darwinian evolution."

"That's one possible explanation—and if it were true, then Scofield would have wasted billions of taxpayer money. But Scofield's skunk works decoded God's recipe. They fully understand the genomic program. Their success is proof, so to speak, of intelligent design. Once the basic mechanisms were decoded by the code breakers, the computer scientists started putting together algorithmic maps of subroutines, and these subroutines were tested by the biologists. As the fabric of the program came together, a larger algorithmic pattern was detected by the code breakers, and the skunk works team was able to see the entire structure at last. This program is very complex, but has been tested and found to be a precise description of the genome operation at the cellular and systemic biological level."

"That's not proof of intelligent design," Mark objected. "Evidence maybe, but not proof. The 'program' could still have evolved through billions of years of Darwinian experimentation. The 'messy' program may be a result, but not a predetermined one."

"Mike has similar doubts," Maria conceded. "I do not. This discovery raises the question: Are we violating God's covenant by discovering the genomic program and tinkering with our bodies, or did God plan this all along? Is self-programmed evolution in God's plan?"

"When you asked Him, what did He say?"

"Behave yourself, pup, or I'll have to spank you!" Maria smiled, releasing the tension between them.

"Each answer Science gives us raises new questions," she continued. "I personally believe that Homo sapiens were embedded in the genomic program from the very beginning, that we were created to give witness to God's wonders in the universe and to become the first sentient beings to reach a level of awareness sufficient to turn our minds inward and begin to program ourselves. Evolution is now in our hands, directly."

"Does Gen. Scofield believe that?"

"He says he doesn't, but I don't believe him. His actions suggest otherwise."

"Isn't this Pandora's box, Maria, the most dangerous invention of all time—one you've unleashed upon an unprepared world?"

"Perhaps. It's rumored that Mike killed everyone in the skunk works project to protect the secret. In any case, they've disappeared. As far as we know, he's the only living person that knows the program, except for one accomplice. This is the one secret he refuses to tell me."

"He's skilled at killing, isn't he? But you said that everyone here is *modified*? If he destroyed the programming team, how has he accomplished that?"

"Mike has created certain *enhancements* to share with his trusted subordinates, and a different set of enhancements for his Templar Knights. He can turn off the aging sequence, for instance—even reverse it, if he wishes.

"He has also created metabolism routines to control weight gain, and muscular enhancements to make us stronger, faster, and quicker. He has enhanced our senses of sight, hearing, taste, and touch. He has improved our healing ability and resistance to viruses and bacteria. Beyond that, he has been cautious. Most of us believe that Mike has made additional adjustments for himself that he isn't sharing. I sometimes think he can read my mind."

CHAPTER THREE
THIRD-PARTY INTEREST

TOUCHING THE TIGER'S TAIL

"I wonder what their objective is—in following so closely? They're bound to be aware of each other's presence, right, Admiral?" Scofield said to Rear Adm. Paul Wainwright, the man in charge of the flagship's counter-submarine warfare operation.

"They probably do have a fix on each other, General. And they're not exactly allies, so it's unlikely that they're coordinating their efforts. Perhaps it's a game of chicken, each sub creeping a bit closer to the tiger until one of them has the guts to pull its tail."

"This tiger bites."

"Yes, sir, it does, as they are about to find out."

"How will you coordinate the attack?"

"We've positioned two sea jets near each submarine. The jets are invisible to the subs and have a standoff of five hundred yards. We've linked their fire control systems to our computers, so that the sea jets will all fire simultaneously. The subs will detect the sea jets' weapons release four seconds after splash, but will have no time to respond or evade. They will be killed ten seconds after launch."

"Will they have time to send out distress signals or notify their commands?"

"Not likely, General. They will be preoccupied with immediate survival, ordering evasive maneuvers. The attack will catch them off guard. It isn't inconceivable, though, that they might have some automatic emergency-alert system."

"How soon will the subs be inoperable?"

"The explosions will be instantaneous and catastrophic—all hands dead within seconds. The subs appear not to have flooded their tubes, and have definitely not opened torpedo tube doors."

"Very well, Admiral. Proceed."

It was go for sinking the four submarines: two Chinese, one Indian, and one Russian. Over five hundred lives would be lost and billions of dollars in equipment sent to the bottom of the ocean. The Russian nuclear ship would cause additional environmental damage.

The sea jets had also come from Scofield's science labs and were as top secret as his other new weapons. They were mini-subs, propelled by a unique ramjet intake system that allowed them to *fly* through the water at over 150 knots. At those speeds, they would have been torn apart by the friction of the water—if not for the nano-armor coating and nano-heat shields, which created a bubble of air around the jets. That bubble also protected the jets against impact with underwater objects. When moving at high speeds, the jets annihilated fish; they were restricted to using those speeds only when tactically necessary.

Amazingly, the jets operated silently up to thirty knots; faster than that, they were pretty noisy. Nano-layers made the jets invisible in daylight, and to sonar and down-looking radar.

More importantly, the sea jets were relatively cheap, almost matching the power of an attack sub for one-thirtieth of the price. Scofield's fleet had over a hundred of these new pack hunters.

The general watched the battle through visual feeds from the eight sea jets arrayed on a set of displays in the command center. The targets were difficult to make out at 500 yards, but the displays were enhanced with light-magnification and image-sharpening programs.

The admiral nodded to one of his officers, who turned a key and pushed a button. Faint trails could be seen on the screen as the small torpedoes left the sea jets and headed toward their targets. The four subs imploded simultaneously.

"Well, that does it, General. We're in the shit now."

"Yep. Good job, men. Let me know if there are any repercussions Admiral."

REPERCUSSIONS

"General!" One of Scofield's aides turned on the light in his stateroom to wake him with news from the command center. Maria, naked from the waist up, with a blanket bunched at her waist, waved at Capt. Tim Weaver as he moved towards the bed.

Weaver smiled, admiring Maria's breasts, then turned his attention to the soundly sleeping general. "Wake up, Mike! You'll want to see this."

Weaver was the general's aide, but also a trusted bodyguard. He was a long-time friend, advisor, and confidant. In private, they called each other by their first names, or worse.

Maria turned over and went back to sleep with a feminine sigh.

Scofield envied her. "Yeah, what is it, Tim?"

"The Chinese are onto our little maneuver. They've accused us of sinking their two submarines and are threatening a counterattack if we don't offer an acceptable explanation. They used the specific language 'an act of war.'"

"They should be careful what they wish for," the general replied groggily. "Okay, give me a minute to clean up, and I'll be right down. Do they have any ships in a threat posture right now?"

"No, sir, but they are moving a small fleet towards the two carrier battle groups that are waiting for us around the horn."

"Have we finished our nano-armoring of those ships yet?"

"No, sir."

"Roger that. Hey, Tim, would you get me some coffee?"

"Do I look like your slave, Mike?"

"I'd really appreciate it, you low-life, ungrateful piece of shit." Scofield loved his Lions, first formed in Afghanistan, of which Tim Weaver was one of the few living original members. Only the snipers had survived the Iranian adventure, and Weaver had been the lead sniper.

"Screw you, old man… Straight up, or latté?"

"Latté, thanks."

The command center was abuzz when Scofield arrived, the admiral agitated. "Mike, I don't know how they found out so quickly," he blurted. "They must have had some warning."

"They probably missed a scheduled check-in. Or maybe they put those subs there expecting us to sink them and justify a fight. They may fear what we might do, and the impact it will have on their oil supplies from the gulf. Knowing that we're currently leaderless, wounded economically, and especially vulnerable, they may think they can flex their muscles and bring us back from the brink of this war." Adm. Wainwright postulated.

"Those Godless communist bastards aren't trying to protect the holy sites; that's for sure." Scofield responded.

"The Arabs may have persuaded them to intervene, or they fear the ramifications of provoking the whole Islamic nation. Their massive nuclear arsenal does throw a checkpoint into our strategy." Wainwright added.

"Capt. Weaver, assemble the Lions. I think we need to pay a personal visit to the General Secretary and help him see the error of his ways. Scofield was now as agitated as Wainwright. "Admiral, when can you get me stealth transportation for our little journey?"

"Give me six hours. We'll require aerial refueling to and from Diego Garcia. Let me make sure we have all the necessary assets. I'll let you know if we can do it faster."

"With that jump-off time, when would I be on the ground in Beijing?"

"Twenty to thirty hours. We'll probably send you from Diego to Iwakuni to prepare for your final approach. I'll put you on a modified Osprey for your flight from Iwakuni, so it will be a slow slog to Beijing." The admiral calculated.

"And a slow slog home. Do you have anything faster? I'd hate to have swim back if the Chinese shoot us down on our return."

"We have some surprises waiting for them if they try."

"How much time do we have before the Chinese can threaten our fleet at Cape Agulhas?"

"Their strike fighters could be in range in three days."

"Okay Admiral, I'll give you five hours till departure. In the meantime, send a message to the Chinese that we are investigating the incident and will give them a formal response in no more than forty-eight hours. Ask their kind indulgence—and ask them, ever so politely, to back the hell off our ships at Agulhas. Tell them if they don't, we'll destroy any ship or aircraft that threatens our navy."

"Roger that. Five hours it is."

"Capt. Weaver, wake me in three hours."

"Yes, sir."

The general walked back to his stateroom, pulled the covers off Maria, and indulged himself, energized by the thought of his imminent adventure. She was happy to oblige, as long as he didn't expect conscious participation.

THE CHINA JAUNT

The mission to China was arduous. It started at 05:30 and ended forty hours later in Beijing. Gen. Scofield, Capt. Weaver, Col. Maria Olsen, and the small group of fifteen warriors handpicked by Weaver from Scofield's Lion's Den were first airlifted from the flagship to the Carrier *Enterprise* on a V-22 Delta Osprey tilt-rotor vertical/short takeoff and landing (VSTOL) aircraft. They launched from the carrier in a modified C-2A Greyhound and flew thirty-five hundred miles to Diego Garcia, exhausting their external fuel tanks en route and refueling mid-air from an aircraft dispatched from Diego. The C-2A had been modified years earlier to allow refueling and to carry external fuel tanks, but was rarely used for this length of a mission.

Adm. Wainwright was greatly relieved that his boss didn't have to try out the amphibious capabilities of his nano-armored suit. He didn't envy the general's group, though. After traveling for over twelve hours to Diego, they would be whisked away without pause on another long leg of their journey.

Diego Garcia is a tiny spit of land in the middle of the Indian Ocean within the Chagos Archipelago. Diego has the primary mission of fleet resupply, pre-provisioning, and a vital land-based airfield—which had been used during both Gulf wars and the wars in Afghanistan and Iran. A British Territory, it's used primarily by the United States, by treaty.

Immediately upon arrival, the group departed Diego in a C-17 Globemaster cargo jet for the Iwakuni Marine Corps Air Station (MCAS) in Japan, a journey of forty-four hundred miles in nine hours, with one mid-air refueling. There, the crew disembarked for some quick chow. It was 02:34 South African time (the time zone of their departure), 09:34 local time. The hop to Beijing would take a little over three hours, with one time-zone change. They would lay over for fourteen hours, to coordinate their arrival with a late-evening surprise the general had planned for the President and General Secretary of China. After resting for eight hours, they rose for hot chow and began the mission briefings for their jaunt to Beijing.

"General, we've established a secure communications link to your flagship," Gunnery Sergeant Bob Saddler interrupted. "Adm. Wainwright is ready to brief you."

"Thank you, Gunny. I'll be along shortly."

Only Maria accompanied the general to the conference room, which was guarded by a small group of Marines in full battle gear. Inside were a simple conference table and an odd "vintage" secure phone attached to a computer, which in turn was connected to a hardened box by a thick cable. The computer was a tempest-capable PC-based computer shielded against electromagnetic radiation emissions to prevent eavesdropping. It was an older, encrypted, and data-compressed communications link, sent via satellite, to Wainwright and the general's assault fleet. The system provided voice, text, and video links between the two commanders, as well as a remote capability to display briefing screens for situation maps or other graphical information.

"Good evening, General," Wainwright began. "How was your nap?"

"Good. Slept like a dead man. It was good to get out of the planes."

"I can imagine. It appears that everything went well."

"Yes. Thank you, Admiral, for all your hard work. The team understands how complicated this mission is for you and appreciates your flawless performance."

"Thank you, General, but let's not jinx this. We still have work to do."

"Roger that. What is the status of the fleet and the operation?"

"The amphibious task force and the USS *Enterprise* carrier strike force have rendezvoused with the USS *Ronald Reagan* carrier task force and are now steaming south of Madagascar. The USS *Nimitz* carrier strike force is deployed ahead of the fleet to the east of Madagascar and is now in range of the Chinese fleet, which sailed perilously close to Diego Garcia. I suppose their actions were meant to intimidate."

"Are you intimidated, Admiral?"

"I wish my fleet was nano-armored, General. We're vulnerable until we are. That won't be completed for another week. We've slowed the fleet down to make sure we get that done before we enter our final theater of operations."

"Roger that. Take the time you need. The ships will be pummeled in the Red Sea, and we need to be protected. Tell Mouse to get the lead out."

"Mouse" held doctorates in computer science, chemistry, and electrical engineering. Mouse was Mike's personal friend and fellow science nerd, and rumored to be the only other person alive who knew the recipe for the "secret sauce." Everyone joked about Mouse, but all knew that the little munchkin was a certifiable genius.

"What's going on in the political arena?" the general asked.

"It's rumored that a loose truce has developed between the hardliners and progressives. We've heard through the four-star commands that they may even have an interim president selected, and that elections will be scheduled for next year."

"We need a new president. It's been weeks since we lost our last one. We should not jump into elections, though. We need stability while we rebuild. I'll talk to both factions when I get done and encourage them to delay elections, if they can. Anything else I should know?"

"Your surprise is ready, General."

"Good. Goodnight, Admiral."

The final leg of their journey was from Iwakuni directly to Beijing, on a specially outfitted Osprey V-22, nano-armored and enhanced for long-distance travel.

This stealthy V-22 aircraft with its VIP strike team would fly, with one mid-air refueling stop just outside Chinese territory, directly to within two miles of the general secretary's lodgings in Beijing. The V-22 could cruise at about 340 miles per hour; the trip to Beijing would take about three hours.

In all, the assault group had been en route for over thirty-nine hours, with a time change of six hours. They arrived at 01:37 Beijing time, landing without incident atop a building selected for its flat, wide roof. The group, in fully cloaked mode, calmly made their way to the General Secretary's residence. For maximum psychological effect, they killed every guard and every attendant in his complex, a time-consuming and bloody task.

Scofield and Maria left this part of the mission to Capt. Weaver and proceeded directly to the Secretary's bedroom. They silently entered his quarters. He was still awake, busily reading documents at his desk, dressed casually in pajamas and a silk robe. The general positioned himself in front of the Secretary's ornate desk and switched off his cloaking mode, leaving his face shield opaque to give himself a frightening appearance. The surprise had the desired effect.

Scofield then turned his face shield mode to visible to let the Secretary see his face. Recovering from the initial shock, the General Secretary caught his breath and tried to compose himself.

Maria left her mask opaque, but the sight of her form-fitting nano-armored suit disturbed the General Secretary as much as the hulking figure of Gen. Scofield. He was amazed at the sight of her. *They send women on such missions?* he thought.

The general turned on his automatic translator and began to speak. "General Secretary, I'm Gen. Michael Benjamin Scofield, United States Marine Corps, at your service. I command the fleet that's steaming, as we speak, toward the Red Sea. I've come to make a full report, as requested by your government, on your two lost submarines." His words were translated to perfect Cantonese.

"I regret to inform you, General Secretary, that your submarines were sunk by my forces forty-eight hours ago off the coast of Africa. I ordered their destruction because they were shadowing my fleet at a time of war—and to send you a message that your country must not impede our mission. I'll repeat that message to you now, in person, if you like." The general gave his speech in a matter-of-fact and professional manner.

"How bold of you to break into my house, General. You won't find it so easy to break out. You have violated international law and precipitated an act of war against my country. That cannot be tolerated."

At that moment, concealed panel doors opened behind the general secretary. Two security agents in black burst forth from their hidden compartments. Bolting aside, Maria drew her nanosword as the two men fired their weapons at her, she moved rapidly to close the distance between them. She quickly dispatched both men, slicing them in two with two vicious strokes of her sword. She then paused to survey the scene, alert, confident, and relaxed.

Switching to intercom mode, the general commented privately to Maria: "You haven't lost it, baby. Just curious: Did any of their bullets strike your armor?"

"Too fast, my love."

Traumatized by the scene before him, the General Secretary, with concerted effort, regained his composure and stilled his shaking body. Maria and Scofield admired his courage and control under duress. *There was a reason this man made it to the top.* Scofield thought.

"I respect your position, General Secretary," Scofield said calmly, "but I regret to tell you that you are in no position to threaten the United States. We're on a war footing; we *will* invade the Arabian nations, and you *will* stand down your forces."

"We won't stand down. You may assume, effective immediately, that a state of war exists between our countries. We regret that your country was attacked by terrorists—we have all suffered from terrorist attacks—but this isn't an excuse to start a world war."

Scofield stared calmly at the General Secretary, marveling at his pluck. Then his face shield went opaque for a full minute, perplexing and infuriating the General Secretary. Clearing his shield at last, Scofield said with unnerving composure, "Call your naval headquarters, General Secretary."

"Why?"

"Call your headquarters, please. The reason will be apparent."

The General Secretary's phone rang just then. He picked it up and began talking with his subordinates. Suddenly, his face turned ashen. "What have you done, General?"

"My headquarters has an active video link to me. They were listening when you declared war moments ago. Your declaration was recorded and taken seriously by the United States. On my command, inaudible to you, our forces sank the Chinese fleet steaming towards our carrier battle group near Madagascar. I believe that you have just been informed of our success in that endeavor."

While en route to Beijing, the general had positioned all of his sea jets around the Chinese fleet, ready to attack simultaneously on command—the command given immediately pursuant to the General Secretary's declaration of war. The Chinese fleet was utterly destroyed. Twelve thousand souls and the first commissioned Chinese aircraft carrier were sent to the deep. No sea jets were lost.

"Do you still wish to make war on the United States, General Secretary? Or would you prefer to withdraw your declaration? The United States would view a withdrawal very favorably."

The General Secretary didn't answer.

"The rest of the Chinese global fleet will be destroyed in precisely three minutes, General Secretary, if you don't act." Scofield bluffed.

"Have you used nuclear weapons, General, to destroy our fleet?" The General Secretary was shaking with rage now.

"We've used only conventional weapons."

"How is that possible?"

"We have *new* conventional weapons, General Secretary."

Struggling to control his fear and rage, the General Secretary responded, "China no longer has an interest in the Red Sea."

"Thank you, General Secretary. I hope this misunderstanding won't affect our countries' relationship in the future."

DOES THE PRESS KNOW?

After his conversation with Maria, Mark Osborne had been given unprecedented access to the general's operations. He was present in the Command Center when the navy sank an entire Chinese fleet, a fleet ready for war and steaming toward the Carrier Battle Groups. Mark couldn't interpret all the commotion, but the body language and banter of the men controlling the battle signaled that the navy was enjoying success.

The officers and enlisted men manning the control screens were clearly excited. "Yeah, baby, get some!" they exclaimed, and "Jesus, did you see that?"

Mark was surprised by their emotion. He imagined that the stakes were very high and that these young men were as scared as they were excited.

"I can tell by the excitement that everything went well," Mark commented to the assistant duty officer, a Navy captain. Mark didn't dare interrupt Adm. Wainwright, whose intense focus, exhaustion, and bad spirits were evident.

"The battle was over faster than we could have hoped. All of the Chinese ships are sunk, or sinking, and the destruction looks total," replied the captain.

"Are you surprised at that?"

"Surprised, no; relieved, yes. In an operation like this, a thousand things can go wrong. If we'd failed, we might have launched World War III, and invited the Chinese to join the other side. Plus, as we were speaking, Gen. Scofield was standing in the private office next to the bedroom of the Chinese General Secretary, threatening him with the action you have seen playing out before you. These sea jets are largely untested, especially after racing at one hundred and ten knots underwater for twenty hours. If anything had gone wrong, it could have turned out vastly different."

"Why did we sink the Chinese fleet?"

"I can't read Scofield's mind, though it's widely rumored that he can read ours, but I can see a number of reasons why he would do it."

"Can you elaborate, Captain?"

"We're exposed here, acting like a wounded animal, lashing out. Moreover, we are about to pick a fight with one-point-three billion people, one fifth of the world's population. It's reasonable to expect that our enemies, our rivals, and even our allies might be disturbed by that prospect. The Chinese were the boldest, and the first to take a poke at us. By crushing them in such a decisive manner, we may have squelched other attempts at impeding our mission. I hope so, anyway."

"If our conventional forces are so dominant on land and sea, won't that push our enemies to use the only weapon they have left, the nuclear option?"

"That's the next threat we must deal with. Luckily, we have a vast and defensible nuclear arsenal as well, so the only takers will be suicidal ones."

MORE POLITICS

"General, the flagship has a priority secure communication request from a Sen. Fitzpatrick," the sergeant said, handing the general a headset. They were just barely into Chinese territorial waters, heading back to Japan.

"That didn't take long. I guess I'm headed to the woodshed," Scofield said.

Maria smiled that smile that said, *Yeah, terrified, I'm sure.*

The general positioned the headset and said, "Scofield here. Go ahead."

"What in holy hell have you done, General?" Fitzpatrick raged.

"We just had a little misunderstanding with the Chinese, Senator. I think we've worked it out."

"I cannot believe that you sank two Chinese submarines, and then an entire Chinese fleet—while standing in front of the goddamn imperial leader of China and threatening him with his life! What could possibly have possessed you to kill every guard, cook, and housekeeper in the General Secretary's household? Hell, General, you killed his fucking mistress."

"Seemed like a good idea. It appears to have worked—Really? His mistress? He must have been pissed at that."

"Have you lost your mind, General? Do you want to declare war on the whole world? I just talked with the General Secretary, and he stopped just short of threatening global nuclear war if you aren't bound and gagged and delivered to him for disembowelment ASAP."

"Yeah, but he didn't mean it. Frankly, I thought we had established a good rapport by the time I left him. He seemed quite relieved not to be at war with us."

"Okay, fine. You think this is a joke and that you are answerable to no one. Do I understand you correctly, General?"

"No, Senator. I assure you, this is not a joke. We are at war, and war requires risks. The Chinese were the first of what would have been many to challenge us. I needed to take dramatic action to quell such insurrections. I think we can be confident that there will be no new takers—at least until they start to understand the technical advantages that allowed us to dominate the Chinese at sea. That caution will be reinforced on the ground when we show them the nano-armored division in action. These demonstrations should buy us time to do what we've set out to do without interference from the rest of the world."

"What should I tell the General Secretary?"

"Reply forcefully, implying that I'm rogue at this time and that you are trying to rein me in. Recommend that the Chinese, or any surrogate they might use, stay clear of my invading force. Tell him I have a substantial nuclear capability under my exclusive control. In other words, tell him the truth."

"I'll do that. Are you familiar with the progress that we are making in selecting an interim president?"

"Yes, I am."

"Then you know that it's highly likely that I'll be your next boss."

"Can't wait." *Asshole*, thought Scofield.

"My first act will be to fire you, General."

"Time will tell, goodbye Senator."

Scofield had reached a point in his life where the dealings of petty men made little impression on him. He knew his role in world affairs and rarely took offense at chuckleheads trying to prick his ego. Still, he suffered from the same tribal tendencies as any man, and had to admit that he truly disliked the senator. He would have to guard himself against those feelings. In the past, he would have crushed a maggot like Fitzpatrick just for sport, but he was more refined now, he thought, smiling to himself at his self-delusion.

Catching Maria studying him, as if wondering what he was thinking, the general hid his smile. He replaced his nano-armored helmet, which automatically mated with the neckline of his uniform. He winked at Maria, then blanked out the screen. From the suit's music and video collection, he selected a rendition of Mozart's "Requiem" and a slide show of his home in Virginia in panoramic wide-screen mode. The slide show he'd created was interspersed with pictures of his children at all ages, his grandchildren, and his wife Sarah.

He listened to the music, allowing himself to feel homesick and to mourn the common man he had been but could be no more. His family no longer knew him; he was lost to them. They followed his exploits, but they could not join him or even understand how their beloved Papaw could do the things he'd been accused of. Remembering with deep longing how much he loved them, then and now, Mike let the music lull him into a deep melancholy, languishing in his sadness. Weeping softly to himself, he initiated the suit's sleep and relaxation program, and within minutes, despite the incredible noise and rough ride of the Osprey, he was blissfully asleep.

Few knew the general's penchant for deep melancholy after close combat and the many deaths he caused directly and indirectly through his command. It was his private ritual, when events allowed, to turn on beautiful music and let it lull him into a sorrowful mood to mourn his victims, his lost comrades, and the distance added between himself and his past.

So much lost. When will I be allowed to rest? he often wondered.

Maria was equal parts excitement and trepidation. These past days had awakened powerful emotions in her, long-buried fear and anger. Despite the rough ride and Mike's conversation with the senator, she had dozed off, waking just in time to see the general wink at her and shut off his facemask. She worried about him at times like these, when his weak smile betrayed his mood. She knew he could become deeply morose after a battle, and that he had never fully recovered from his wounds in Afghanistan and Iran. Despite the genetic magic he used to heal his physical body, she knew a part of him longed for an end to his waking nightmare. He was too strong to succumb to such thoughts, but they haunted him nonetheless. Many of the personal risks he took seemed to be challenges to the Fates. "Go ahead, kill me, you bastards!" became his unspoken mantra as he looked death in the eye, time and again. This had never been truer than in the campaign in Iran following his Afghanistan battles— he'd taken crazy risks then. This night, her only course was to leave him alone and let him heal.

Maria dozed again. She awoke disoriented, thinking she was back at the convent. She looked around, panicked, but quickly realized where she was and what was happening.

Maria turned on her pleasure enhancement routine for a little distraction, releasing her body to the touch of a billion pleasure-directed specialized nano-bots, whose sole purpose was her gratification. The thought of Scofield and Mouse making an exception for her, and leaving the masturbation layer in her suit, made her smile. Those two were wicked indeed—and probably watching though some fancy recording routine. She felt more than a tinge of guilt as she thought about the last two weeks and how she had allowed her sexual self to reawaken. *Sex is part of God's natural order* she rationalized, feebly. She supposed that it must be the heightened tension and excitement of returning to combat, and seeing all her boys again, that had rekindled her once legendary and unquenchable lust. *How can I let two years of dedication to God vanish so easily? What is wrong with me? Forgive my sin Lord.*

Maria noticed Capt. Weaver watching her with amusement. She smiled at him and mouthed the words, "Be nice," then turned her mask to opaque. *I wonder what the general is doing,* she thought, as she ordered up the sleep routine and drifted off, sated, and happy to be again by the side of her one true love. *This place is my destiny,* she thought drowsily. *I'll die by his side one day, maybe soon.*

A Marine sergeant, the plane's crew chief, observed the oddly relaxed behavior of his VIP guests. He was perplexed that men and women who had just slaughtered so many innocents could appear so unaffected by their actions. He had been in combat in Iran and had seen the carnage unleashed by the Marine Corps after their regiment was wiped out by the Iranian incursion into Afghanistan. That war had been led by this revered Marine, Gen. Scofield. Still, he remembered feeling physical revulsion at the slaughter of all those Iranians, many of them young conscript boys. He never got used to it. *These people are different from the rest of us,* he thought. *Sometimes you have to kill, but you don't have to like it.*

BACK AT JAPAN

The Osprey arrived without incident in Iwakuni at 08:30 local time. The general ordered his team to stand down. They might take another side trip before going back to the ship, he confided, so they should run maintenance, get some sleep, and make ready to move.

Maria accompanied the general to the secure temporary conference room to be briefed by Adm. Wainwright. Both were refreshed after their naps, but Wainwright looked exhausted. The general made a mental note to order Wainwright, on whom Scofield was increasingly dependent, to get a solid eight hours of drug-induced sleep.

"What has been going on, Admiral?" Scofield asked.

"Everything has been relatively quiet, General. As you know, the extraction from Chinese territory went without incident, and we've seen no signs of other Chinese or non-Arabic naval resources moving into our area of operations. The *Nimitz* Carrier Battle Group has slowed to allow the rest of the fleet to catch up. In a few days, no more than a week, we should be ready to start our final combined and coordinated approach toward the Red Sea.

"There has been substantial Arabic naval, land and sea activity near the Red Sea, with a growing capability converging around Yemen. They may try to stop us before we enter the sea, at least with some of their naval forces. Another group of naval resources, submarines included, is gathering near Jeddah. They no doubt expect to defend both locations, given our superiority and the likelihood that their first force will be defeated. Those forces are probably meant to bleed us before we enter the Red Sea.

"We're also seeing a convergence of air assets. Every available airfield is being requisitioned for military use by every nation in the area, including Sudan. Ground forces are moving into the operational area from every country of significance: Turkey, Egypt, Libya, Sudan, Indonesia, Syria, Jordan, Saudi Arabia, Iraq, Iran, Pakistan, Afghanistan, Yemen, Oman, the United Arab Emirates, and Kuwait are all sending forces towards Mecca. These forces are at different stages of deployment and vary greatly in size and composition. If all converge, their combined force may exceed two or three million men.

Additionally, Turkey has asked NATO to stand by them and help thwart this invasion.

"I guess we are having the desired effect," the general mused. "Feels like Thermopylae. I hope our outcome is better."

"Leonidas only had three hundred Spartans."

"Yeah, but he was backed by six thousand free Greeks in the rear."

"The well-equipped and smartly trained Spartan Hoplites are the proper metaphor for nano-armored Marines, and we are ten-thousand strong." Wainwright added.

"True. We're only outnumbered two hundred to one, good odds for any Marine," said Scofield, lifting lines from the movie *300*. Both men laughed. "What else?"

"The Pakistani operation is nearly ready. If you want to participate, you should start moving back to Diego."

"What about Operation Rathole?"

"We're working on that. I suggest executing the Pakistan operation first to eliminate the immediate nuclear threat, and going after the leadership group later as we start the naval preparation for the invasion."

"Roger that, Admiral. One more thing: Have Maj. Wayne Clark call me from an isolation/encryption communications room as soon as you can."

FORMING A NEW GOVERNMENT

"General, this is Maj. Clark. What can I do for you?"

"I have an operation that I want you to run for me. Let's encapsulate it in deep-compartmented mode, need-to-know-only, and only if cleared through me."

"Yes, sir."

"I want you to select two officers from the teams we left behind in the States that have extraordinary good sense. I want officers who are forceful but respectful to a fault."

"Yes, sir."

71

"The new provisional Congress is meeting in Philadelphia soon. I want these two officers to infiltrate the proceedings of the provisional Senate and be ready to report back to me on command. They're to take no action, and are not to be seen or discovered without my expressed permission."

"Yes, sir."

"When these men are in place, I want you to join them from the flagship on a secure communications link to me as soon as possible. Tell them that the situation will likely be fluid and that they must remain flexible."

COMPROMISE

"Do we have a deal, Sen. Fitzpatrick?" Gov. Raymond Jones was exhausted from all the back-and-forth negotiations.

"Yes, Governor, we do."

"By my count, we have seventeen new senators appointed thus far by governors throughout the country. That leaves seventy-six unfilled seats. You have agreed to seat and swear in all of the current appointees by Monday of next week." Talking over an encrypted video link provided by the military, the governor recapped the deal one more time.

"That's correct, Governor. Subsequent appointees will be sworn in after the special session to elect the interim president."

"That will leave seventy-six votes by governors whose states have at least one remaining Senate vacancy. This vote will be cast specifically for electing the next President and nothing else. Did I state that correctly?" The governors had fought hard for this concession.

"Yes you did, Governor. We'll vote on the interim president no later than Wednesday of the following week when we convene in Philadelphia."

This was a tough concession for Fitzpatrick, but he had few options. He knew that support for unilateral congressional action was losing ground. He was also beginning to believe that the progressive group he now firmly led could marshal the fifty-one votes required to vote him in. One Senate appointee had already optioned his vote secretly to Fitzpatrick to gain a leadership position in the new Senate. This new senator had been appointed by the hardliner-leaning governor from Missouri and counted as a solid steal from the other side. With forty-seven votes locked up, Fitzpatrick needed to flip only four more ambitious new senators. He had pushed hard for more appointees to the Senate—targets for his pork-barreled gun—but appointments had stopped when the governors learned that they could vote instead.

Governor Jones was pleased with his little ruse. He had sent his "double-agent" senate appointee to Fitzpatrick, to convince him that he might win after all—and thus persuade him to assent more readily to the governors' voting plan. Jones was worried that Fitzpatrick would drag the process out forever, and they didn't have forever. Gen. Scofield had just wiped out the Chinese fleet, and God only knew what he would do next. Raymond Jones needed to be president soon, so that he could "support" Scofield and claim the general's glory as his own. With a successful campaign against the Arabs under his belt, he would be elected president in his own right in the special election, erasing the asterisk behind his name in the history books. Scofield had better not screw up.

"I'll inform the senators."

"I'll inform the governors."

Capt. Dick Riley was standing in the room with Gov. Ray Jones, listening to the exchange. His counterpart, Capt. Jim Deissler, was on the other end with Fitzpatrick. The underhanded dealings of the two politicians had removed any qualms they might have had about spying on them—an espionage effort that would continue right up to next week's meeting in Philadelphia. To prepare a briefing for Gen. Scofield, the two captains commanded the artificial intelligence (AI) engines in their suits to record and transcribe the event they'd just witnessed. The narratives would be automatically hypertext-annotated with video and audio and sent to Maj. Clark without review. Clark would merge the two reports for the general.

FITZPATRICK'S LAIR

"**O**kay, we know we can count on, without equivocation, forty-seven votes, right?"

"At least that many, Senator," said Greg Bokert, Sen. Fitzpatrick's chief of staff for legislative affairs. Bokert had been traveling with the senator when the bomb went off, but his family had been wiped out in the firestorm that had consumed his Old Town Alexandria home. Embittered by his loss, he had secretly committed himself to avenging the deaths of his beautiful babies and their mother and, unbeknownst to Fitzpatrick, was collaborating with Jones. The actual number of committed senators was thirty-six.

"Good," Fitzpatrick said. "You folks will work around the clock to solidify the forty-seven votes that are in our camp, and I'll work on the next four we need for victory. Greg, I want you to prioritize the other sixteen senate appointees according to 'convertibility.'"

"Yes, sir."

Bokert would prioritize the list in three groupings: Jones' six additional turncoat recruits at the top, the unconvertables next, and the likely flippers at the bottom. He would arrange the list carefully to avoid detection, but the six guaranteed "successes" at the top would likely cause Fitzpatrick to lose interest in the rest. He had stressed to Jones that each appointee must bargain hard for money or position in the Senate, to keep Fitzpatrick from growing suspicious. And the longer they took to turn, the less time Fitzpatrick would have to turn the others.

Capt. Jim Deissler surveyed the scene in front of him, setting his AI engine to analyze the expressions and behavior of each man. It detected two anomalies: one in the senator, one in Bokert. The program could not determine if the senator was lying, telling the truth, or merely befuddled. It *had* concluded, however, that Greg Bokert was concealing some kind of inner stress, that he was probably withholding information, and more than likely was lying to Fitzpatrick. The AI program had tagged Bokert's report with information on the gruesome fate of his family just weeks before. Capt. Deissler decided to follow Bokert that night instead of Fitzpatrick.

A DEBT DUE

"What do you want, Greg, when this is all over?" Frank Bouchet asked Greg Bokert in a Chicago park near Bokert's house. Bouchet was Bokert's counterpart in Gov. Blake Tilden's office. Their jobs, though similar in title, were worlds apart. A governor's chief of staff had to deal with the myriad daily crises that beset an operating state government with its billion-dollar budgets. A legislative aide was more like a big-league babysitter. Bokert's life was filled with inane tasks meaningful only to an egotistical senator. Bouchet was a doer; Bokert a sycophant.

"I want to be there when Scofield slaughters those bastards."

"Those boys play for keeps, Greg. We should let them do their jobs and not interfere."

"I want to be there when Scofield executes those bastards." His gaze hardened.

"Well, I can't promise that, but I promise I'll try. We're grateful for your help, and your patriotism may yet save this country. If there is a way, I'll find it. Meanwhile, Greg, you should start doing pushups and jogging daily if you want to survive more than five seconds on those killing fields."

"I will. Thanks, Frank."

The two exchanged identical briefcases and walked away.

Capt. Deissler again consulted his AI program for anomalous behavior, but the program found none. Both men were telling the truth. Deissler sent a special report to Maj. Clark, highlighting the day's events, ending with Greg Bokert's desire to fight their shared enemies. He attached a request for permission to mentor Bokert, prepare him for nanosuit combat, and have him assigned to a line unit under his command. The request was likely to be rejected. If not, though, Deissler knew that this could be his ticket to the front.

Amazingly, less than thirty minutes later, Deissler received his answer—not from Maj. Clark, but from Gen. Scofield himself: REQUEST TO TRAIN BOKERT APPROVED. LEVEL-2 GENE PROGRAM. WILL SEND SUIT. CAPTS. DEISSLER, RILEY, AND PVT BOKERT WILL BE TRANSPORTED TO BATTLEFIELD. MEANTIME MAINTAIN POST. SCOFIELD OUT.

Greg Bokert made his way back to his hotel room alone. He had told no one about the phone call he had received the day of the attack. He scarcely believed it, not sure if it was a dream. The memory was seared onto his soul.

When he'd seen the news on the day of the explosions, he had tried to get through to his wife; he'd prayed that maybe they were just far enough away from the epicenter to survive. Each attempt was met with busy signals. Then miraculously, Evelyn had reached him on his cell phone.

"Greg, oh my God, I'm so glad I got you. I've been trying so hard to reach you. My cell phone is completely dead."

"Are you okay, Ev? How are the babies?"

"We're trapped Greg. There's no place to go." Greg could hear the roar of the firestorm through the mouthpiece of their landline phone.

"No! No matter what, don't give up. You have to get out Ev. You have to try." Greg was near panic, tears streaming down his face as the weight of his helplessness closed in on him.

"The fire is everywhere, Greg. We tried to get out, but it was like walking into a furnace; the girls and I have moved into the basement, and I've doused a bunch of blankets with water to keep us safe. I don't know what else to do." A deep, hoarse cough resounded as she said this. Greg also heard coughs from his baby girls, ages two and four.

"I can't lose you, Evelyn! I can't live without you."

"I'm so afraid Greg. What can I do?" The coughing was getting worse from all three.

"I don't know baby. Can you get to some better air? Can you get lower on the floor?" Evelyn didn't respond to his question. "Ev, can you hear me?"

Only coughs and desperate gasps emanated from the phone. "I love you baby." Greg said softly. Then, finally, silence.

The roar from the fire increased—reaching out over a thousand miles of fiber-optic cable to the cell tower connecting his phone to his family, mocking him. He could not stop listening, hoping for one last word from his family. It was then that he heard his daughter Marcia's manic screams. She was on fire. Then the phone went dead.

COALESCENCE

"Victory is assured for a candidate from our side. I know you want the job, but a consensus has not yet formed," said Governor Tilden, briefing Governor Jones on the rough vote tallies.

"If we fight amongst ourselves, it's possible one of the others will be elected," Governor Jones protested. Tilden had become the power broker in this high-stakes poker game. Those who considered Jones too radical had rallied around Tilden.

"Doubtful, Ray. We all agreed that the winner had to get a fifty-one vote majority. This was a departure from the tradition established by the founders when they elected George Washington—as was the choice of the Senate over the House for electing the President, which is in direct violation of the Constitution. Then, there's this business about letting the governors vote, instead of the senators they should've appointed. But our voting block is solid and won't easily crumble."

"You never know in politics. It would be better to close this in a one-ballot victory and shut down any ray of hope for Fitzpatrick."

"Well, if you mean that, then we should both step aside and elect someone acceptable to the group as a whole."

"Are you willing to do that Blake?"

"For the right guy, yes, I would gladly step aside."

"Who's the right guy?"

"I would throw out Stanley Donner's name gladly. If you and I can support him, it's a done deal."

Stanley Donner was the rare politician, patriot, historian, and leader that men of all stripes admired. He had been appointed acting Secretary of State following the death of his boss in the DC blast. He wasn't the down-and-dirty brawler who could win a national election, but he *was* the perfect choice for the perilous times at hand. His being a former Marine, having served and been wounded under Scofield in Afghanistan and Iran, might give him the chops to rein in the general. Scofield would not respond to coercion, but he might follow this man's lead.

Jones hesitated, pondering the implications of Donner as the next president, and its impact on his lifelong ambition. "Country and honor before thyself," he said, at last. "I agree."

"Country and honor before thyself," Tilden replied.

Capt. Riley relayed the news to Scofield in a flash priority message from his suit via encrypted satellite link. In ten minutes, he got his response: EXCELLENT NEWS. STAY WITH ASSIGNED PARTY UNTIL DEED IS DONE. MEET UP WITH DEISSLER AND PVT BOKERT FOR TRAVEL TO BATTLEFIELD. GOOD WORK. SCOFIELD OUT.

Deissler had told Riley that they would be transported to the battlefield, but he didn't dare believe it until he saw Scofield's flash response. Riley was ecstatic to be rejoining his battalion and fighting his enemies. Like Greg Bokert, he too had suffered from the DC bombing. His fiancée had been burned in the blast, and her survival was still uncertain. If she lived, she would be horribly disfigured. Riley was a decorated combat veteran who also wanted revenge. Scofield was providing that opportunity.

The events of the last few days had shaken Riley's faith in his country's system of government and its leaders. Today, two men—two politicians—had helped to restore that faith. Still, Riley preferred the Marine Corps, where assholes were visible and easily dealt with. Politicians seem to lose the capacity to recognize truth. The game became more important for them than the outcome.

TO BECOME A TEMPLAR KNIGHT

"Becoming a Templar Knight in Gen. Scofield's division has certain ramifications," Capt. Deissler explained to Bokert. "You'll need to accept major changes, some of which will alter your physical nature in appealing ways. Others will require sacrifices that you may not be willing, or trusting enough, to accept."

"Why are they called *Templar Knights?*" Bokert asked.

"Two reasons, I guess: first, to piss off the ragheads; second, to accentuate the historical parallels between the Templar Knights and us. The division is composed of Christian men and women only; there is no diversity in our ranks.

The Templar Knights of the Middle Ages were sworn to the service of God, taking vows of poverty, chastity, and obedience. We have different views on chastity these days but you'll be asked to agree to similar sacrifices."

"Are you serious?"

"Deadly serious. Our officers and noncommissioned officers (NCOs) were drawn from the Marine Corps, but most of the enlisted men were recruited by special teams using strict mental and emotional criteria."

"Tell me the good parts first."

"You'll be given a 'genetic elixir' that will rapidly initiate physical changes in your body. You'll lose your considerable baby fat, gain muscle mass, and develop stronger, more flexible joints. Your eyesight will become sharper, your hearing more acute, and your reflexes lightning fast. Any lingering injuries will disappear. You'll find that you rarely get sick and that you heal amazingly fast when nicked. Your immune system and metabolism will operate on overdrive. You'll eat ravenously, without gaining weight, and sleep four hours or less each night without feeling tired. In short, you'll feel like a fourteen-year-old on steroids. You'll also be given some optional enhancements for your pleasure, to compensate for the things you must give up."

"What optional changes?"

"An enhanced weapon between your legs, if you like, with enhanced pleasure centers to match it."

"So, the rumor of Gen. Scofield's genetic tampering is true?"

"We don't ask questions, and the general doesn't answer them."

"Does anyone *not* choose the optional enhancements?"

"Not that I know of. After hearing of the sacrifices they're required to make, most recruits gladly accept the enhancements."

"Tell me about the down side."

"The process begins with a surgical orchiectomy."

"What's an orchiectomy?"

"It's the technical term we use to keep recruits from fainting. It means castration."

"You're kidding, right?"

"No. We all go through it. It must be done at least three to four days before the first treatment to insure that no sperm have your new genetic code attached. The general's rules require that any genetic modifications be done to vanilla human genome strands. He won't allow a second-generation human mutation from a modified human. This keeps the baseline human genome completely under nature's control and prevents engineered mutations from getting into the gene pool.

"You are tested to make sure you are completely sterile before we proceed with the good stuff."

"I don't know, Capt. Deissler. That sounds pretty extreme."

"That's what I said, too. It *is* a weird experience, and it takes about three weeks for the enhancements to take effect, so you're out of commission for over four weeks—not that you're able to care."

"But how can you *function* after that?"

"The testes only have two real purposes: to produce testosterone and to produce sperm. The enhancements you receive replace testosterone with a hormone better suited to modern life. Your ejaculate is produced in the prostrate, which is enhanced for pleasure and ejaculate quantity. So, the only thing you don't have is sperm, which you would lose even in a vasectomy. In any case, we can't do a vasectomy, since the sperm are still produced, just blocked from the ejaculate; it must be an orchiectomy."

"What is it like *afterward*?"

"Your sex drive will be diminished somewhat, with your testosterone replaced by a 'kinder, gentler' hormone that makes sex more of an appetite than a craving. It's quite nice, actually. It puts your brain back in command of your member, so to speak. When you do have sex, enhanced pleasure centers more than compensate for any diminished sex drive. Sex is highly evolved, post modification. Orgasms are intense—more intense than you have ever felt, and you can go on forever."

"What does it look like after the surgery?"

"The elixir regrows inert testes to take the place of the originals. It makes the transition easier for most men, so you look the same as before, only better."

"How is it done?"

"Surgically, under local anesthesia—it's really no big deal. You won't feel a thing."

"Wow. That's a big decision."

"Yes, and normally, we insist that you take at least a week to think it over. But you have to decide now, Greg. We have very little time before combat. As it is, you'll be very poorly trained, compared to the rest of my men."

"Tonight?"

"Tonight. Many of the Templar Knights have their sperm and eggs frozen before treatment in case they want children later."

"I could never replace my beautiful little girls. I could never feel that pain again." Bokert paused for a long time, then said, "Okay, I'll do it."

"One more thing Greg—and this is important." Deissler looked somber. "After the nurses have tested you to make sure you are sterile, the next step in your transition isn't so easy. In fact, about one in ten die from the experience."

"What do you mean?"

"The injection of the genetic elixir is extraordinarily painful. We don't give recruits any support or pain medication; it's a rite of passage, and the pain is part of the transformation process. Of those who die, eighty percent are suicides."

"It's that bad?"

"Actually, it's much worse than you can imagine. Only those with faith endure."

Greg paused to process this new information. *Can't be less than my babies endured. If I die, then God will take me to them...* he thought, then blurted, "I'll do it."

"Kneel, Gregory Adam Bokert," Deissler commanded.

Bokert obeyed.

"Will you swear to defend with your life, with all your strength, and by your speech, the holy doctrines of the Trinity and the Christian faith?"

"I will."

"Will you promise to obey the general who leads you, and to defend your country against the Infidels?"

"I will."

"Will you, with your right hand and sword, dedicate yourself to the service of the United States and your faith against the Islamic infidels?"

"I will."

"Rise then, Templar Knight recruit. Strength and honor!"

Deissler explained that whenever an officer said, "Strength and honor," an enlisted Knight would respond with a heartfelt "And a good death!" Usually proclaimed just before battle, it was a reminder of their holy Templar oath borrowed from antiquity, and an affirmation of the bond among Templars facing shared danger.

"And a good death!" Bokert repeated.

"Nurses Furlong and DuPont will take you into the next room to perform the surgery." Both nurses came into the room, smiled and extended their hands to signal that Greg should follow them.

"I'm Sally Furlong," said one, "and this is Mary DuPont. We'll be your nurses during your healing and transformation. We're both Templar Knights and part of the sworn elite that tend to holy recruits. We'll take good care of you, Templar-recruit Bokert."

Both women were accomplished trauma surgeons. They'd undergone their transformations when the general was first building his army. They had been with him every step of the way and had helped most of the Templar Knights' transitions. Sally and Mary knew, better than most, how this initiation process could transform a man. Both suppressed the fact that they were doctors, preferring the intimacy of nursing. Both had forsaken the secular world.

When the surgery was done, Bokert was feeling disoriented at the thought of what had just happened and wondering if he had done the right thing.

"Is it always done this way?" he asked.

"Yes. It's symbolic of the pain of rebirth and part of your initiation. It's not likely that you'll forget that moment, or your sacred oath." Mary said.

"I see that now," Bokert said.

WORKING THEIR WAY BACK

Gen. Scofield had yet to decide whether he would participate in the Pakistan incursion, but it made sense to return to Diego Garcia either way. He was concerned about the delicate negotiations in Philadelphia and needed to be close to the fleet, ready for the naval battle brewing. But his buccaneer nature was calling him to Pakistan, to witness firsthand the fear in the eyes of its leaders as they met their just punishment. Neutralizing the Pakistani nuclear arsenal was also vitally important. He knew his teams could handle it, however, and that he was needed elsewhere.

"We'll embark in fifteen minutes," the general replied to the gunnery sergeant tasked with boarding the VIP guests safely and on time.

The general surveyed his team, letting his eyes linger on Maria. He could read people's moods—their minds, so to speak—in a simplistic way, by sensing covert emotions. It was an imperfect skill that his genetic programmer was trying to improve for him. Tampering with the code to improve brain function was tricky stuff. Scofield was the only modified human who had this alteration, and so it would remain as long he could control the process. It gave him an empathic link to anyone he probed, allowing him to detect such things as dishonesty, sadness, confusion, fear, and embarrassment. This skill helped him listen more attentively to people communicating with him, and understand them better. In Maria, he sensed a deep, otherworldly love and devotion—which endeared her to him almost painfully, so acute was his reciprocal love and devotion. He wanted to put Maria in a safe place to protect her, but knew that she would never allow it. Whenever he was in harm's way, she would stand with him.

Maria looked up, returning his gaze and smiling one of her angelic smiles. She sensed his thoughts.

The team boarded the C-5B Galaxy for the long flight to Diego Garcia. Maria sat down next to the general. "Osborne wants to meet us at Diego," she began. "He says he wants to be where the action is, but I think he really misses me more than you. You okay with that?"

"Is he operationally sound in his suit?"

"He's been checked out. His suit is a noncombatant version, so there is less to learn. Diego Garcia is the primary support base for the fleet now that it's fully deployed in the Indian Ocean. There are a lot of flights daily to and from the base, so he won't be a burden on resources."

"Fine. I can't promise that he can come along to Pakistan, but he can hang out with us in the meantime, if he likes. It will give us time to talk."

"So you have decided to go to Pakistan for the nuke operation?"

The general frowned. He knew she was trying to goad him.

"So, are we going or not? If so, I'll need time to get my nails done and hair fixed."

Scofield ignored the jibe.

"It has been a long time since we last visited the 'land of the pure,'" Maria continued. "I don't have fond memories of our last trip." She immediately regretted her words. Years before, then-Col. Scofield and Staff Sargeant Olsen had led an incursion into Pakistani territory to hunt Osama Bin Laden. They had lost a dear friend and trusted warrior. The Iranian invasion of Afghanistan and the demise of the Third Marine Regiment had followed. It was a memory neither cared to conjure.

"I haven't actually decided, but I'm leaning towards going."

"Sorry about that."

"Don't be, Maria. I haven't thought about Adrian for a long time. I need to remember the fallen; they preserve my past, and caution my future. I'll let you know as soon as I know."

The flight would be long and tedious. The general made all team members wear their nano-armor suits while traveling, in case of attack, and most of the warriors kept their helmets on to take advantage of the suit's restful or diverting programs. Putting his on, the general ordered up combat videos from the Afghanistan operation in which Adrian Lockley was killed. He remembered Adrian for his warm laugh, his stupid jokes, and his staunch loyalty. Adrian had been a fine warrior, one of the original Lions. *One of the dead now...so many dead.*

Why am I spending so much time remembering the dead? the general wondered. He was under stress, of course, but was he wallowing in depression?

Sadness could be hypnotic, he knew, and addictive. He would have to guard against it. What he was unwilling to admit, however, was his uncertainty. Was he doing the right thing, prosecuting this war so aggressively? Should he pull back? Should he let Donner inject himself into the debate? Mike Scofield wasn't one for second-guessing himself, but this operation held the lives of tens of millions in a precarious balance, and he was the one, ultimately, who would decide who lived and who died. Punishing himself with ghosts from the past kept his anger grounded and the cost of his actions in focus.

Pondering these thoughts, he grew more relaxed; almost nodding off, when suddenly he started. Looking up, he saw the Archangel Michael, the Angel of the Army of God, materialize. Scofield wasn't alarmed; he had been visited before, but he became immediately alert. His face screen seemed to dissolve as Michael stood before him. As in a dream, the rest of the plane's interior dissolved, and they were each suspended in an opaque ether.

The Archangel's lips didn't move, but Scofield heard his words: "It is good you suffer, Michael Ben Scofield. You are a bearer of death."

"Why do you bring me further anguish then?"

"I bring you news. The man you seek is upon this earth, born thirty years ago this day."

"Am I to slay this man, as prophesied?"

"He calls forth this conflagration; he orchestrates the actions that cause you to be a destroyer of nations, but you shall not be his slayer."

"Who shall slay him, then?"

"Maria shall slay this usurper, to avenge your death. She's the earthly angel who will restore balance."

Scofield awakened with a start, jumping up from his seat, poised for action. Isolated in their suits, the rest of the team took no notice—except for Capt. Weaver, monitoring Scofield with a programmed avatar. When the general jumped, Weaver was at the ready, but Scofield waved him down.

Scofield wondered if these visions—so real, so vivid, so well remembered—might be illusory. If so, he was clearly delusional, and destined for madness. If not, his crusade might have divine warrant, and be beyond his control. Chosen or crazy, there was no way to know. But he was a Marine at heart; he would leave psychiatric evaluation to the Fates and turn to the trinity of the Corps: duty, honor, country.

CLOSING THE DEAL

"You look like hell, Greg," Sen. Fitzpatrick said as Bokert arrived uncharacteristically late to the meeting. "What happened? Stay up too late fornicating?"

The senator's joke seemed horribly inappropriate to a man still grieving the loss of his family. Bokert forced a weak smile, thinking to himself, *You miserable bastard. I cannot wait to see your face when I screw you next week.* "I slipped on some water when I got out of the tub and strained my hip."

"That would be why you're moving so slowly, I guess."

"Yeah, I guess so."

"Okay, let's get to work. The list you gave me is panning out quite nicely. I worry, though, that I won't have any cabinet positions or ambassadorships left after we get through this," said Fitzpatrick, laughing at his own joke.

"Keep plugging away, Senator. What's our count?"

"So far, five for five. Each step down the list it seems to get harder. Of course, we may not need any more. By my count, we stand at fifty-two votes. Given the expense associated with buying these votes, I may want to stop and save some bullets for later."

"I'd like to see at least two more votes wrapped up, Senator. If these guys will turn for us, they'll turn for Jones too, for a better price."

Bokert wanted the senator to remain preoccupied for another two days. If he reached the seventh candidate on the list, he would stall there for sure. Senator-designate Roland Johns would let Fitzpatrick woo him until the last minute, but would never turn. Given Fitzpatrick's confidence level and his inherent laziness, he would be sure to give up the effort, with no time left for fretting. Some of the selectees on the bottom of the list might actually turn, if approached, and Bokert didn't want him to get that far.

"It's such a pain in the ass, Greg. We have enough."

"Don't get lazy on me now, Senator. You're the only one who can do this, and we need to make sure."

"Okay, fine. When do we fly to Philly?"

"I think we should stay with this process until we're sure of success. After all, the Presidency is at stake. I have a private plane waiting to take us to Philly Sunday night." Bokert responded.

"Okay. Let me know when to be ready."

"We won't leave you behind, Senator."

The meeting continued for four more hours, but, finally, Greg was able to extricate himself.

When he returned to his room, the nurses went to work on him and quickly relieved his pain. They drew a bath for him, which made him feel refreshed and relaxed after his long day. After drying himself off they motioned for him to lie on the bed and, to his mortification, they attached a machine to his penis.

To his surprise, they elicited an erection. "I thought you couldn't get an erection after being castrated?"

"Not true. You still have lots of testosterone flowing though your system. It takes days to drain. You may have some residual sperm left in your tubes that we have to make sure is expunged. The actual mechanics of getting an erection are unaffected. You'll even be able to ejaculate. See?"

If the machine didn't mortify him, this certainly did. Bokert turned crimson. "So, what good did it do to castrate harem guards?"

"The mechanics are still there, but without testosterone, a man loses his desire within about four days. Since harem guards had no seed, they posed little threat to the Caliph's breeding rights. In some harems, of course, the innermost circle of guards also had penectomies, just to make sure."

"Okay, that's enough history, thanks," Bokert cringed. The nurses laughed at his discomfort.

"Doesn't anything embarrass you guys?" Bokert asked.

"Not much," Mary DuPont responded for both of them. They looked at each other and chuckled.

DIEGO GARCIA

"Mark, nice to see you," Gen. Scofield greeted Mark Osborne while he deplaned.

"Hello, General, Col. Olsen, Capt. Weaver. How're y'all doing? Long trip?" Seeing these people in this relaxed atmosphere, knowing where they had just come from and what they had done, made Osborne uneasy. Their smooth transition between the world of the living and the world of the dead and dying was disconcerting.

"Good to see you, Mark," Maria said. "When did you get in?"

"Last night. I'm rested and ready to go."

"Capt. Weaver, arrange a secure meeting room and have the plasma displays set up with a link from Wainwright," Gen. Scofield ordered, shifting effortlessly to command mode.

"Already done, General. They're waiting for us. We also have two command center controllers from the flagship to man the consoles and computers. The systems are secure and robust enough to handle the war from here for as long as you stay at Diego."

"Good. Is Wainwright ready to brief us?"

"The top brass are ready, sir. Adm. Franklin will lead the brief today. They're ready to engage the naval resources arrayed against us."

"Adm. Franklin, nice to see you," Scofield said as he entered the meeting room. "You look tired, old friend."

"The Navy is on duty twenty-four/seven, unlike you lazy Marines, sleeping late and fornicating all day."

"True, the life of a Marine aboard ship is a lazy one. Good thing we have squids to take care of us. Of course, we make up for our slovenly behavior once we hit the beach."

"This time will be no different, I fear. We have a lot of things popping, General, and we need your input."

"Roger that. Go ahead."

"The operational situation in the Indian Ocean is coming into focus. Our forces are well positioned, about four hundred miles northwest of Diego Garcia.

We've sterilized the area in preparation for our first naval conflict since the Chinese encounter."

"Good work on that operation, Admiral."

"Thanks, Mike. The fleet was pleased to receive your message of congratulations. Diego is operating as our forward supply base, with the Air Force providing over watch from there. They're keeping B-52, B-1, and B-2 flights aloft, armed with anti-ship and anti-air missiles past our range of operations from the carriers. We've warned all nations of the world that this is a war zone and to steer clear. Entering this area of operations will be considered an act of war, and the offending ship, ships, or fleet sunk."

"How is that being received?"

"Not well, General, not well. I've stopped answering the phone and turned the job of interfacing with foreign powers over to the Defense and State Departments. So far, we have only two non-Islamic threats on the horizon. The Russians are maintaining a fleet off the coast of India, and the Indians are maneuvering farther north. None are within the restricted zone, but could threaten us within two or three days' time if they decided to do so."

"What is the status of our satellites?"

Franklin continued: "No problems yet. It has been hands-off so far. We have fully deployed our required satellite systems to our theater of operations and have a few spares ready if needed. Attacks on our satellites are always a risk and can put us in crisis in a matter of minutes. We're ready with low-earth airborne backups from Diego, Europe, and Japan if we are blinded. An attack on our satellites is a genuine threat. However, satellite warfare is a risky business. Everybody has satellites and all are vulnerable. Shooting one down invites retaliation. This provides a certain balance of power, and that has kept satellites off limits for a long time. Current intelligence estimates indicate that no Islamic nation is armed with anti-satellite missiles."

"Are the ships nano-armored yet?" Scofield queried.

"Your mad scientist just finished that job. I believe Mouse is vectoring to your position as we speak."

"Did Mouse behave as your guest, Admiral?"

"No, of course not, out of control, like always. We were showed a demonstration of the armor 'paint' on our ships, though. Mouse may be a pain in the ass, but God bless that little shit, that stuff is amazing!"

"Tell me about it."

"Mouse tried to explain the details, but seeing it in action was more powerful than any explanation. The armor is always turned 'on,' but it can be penetrated by a large ship-to-ship explosive or thermal missile if not placed into 'active' protect mode. We would leave it in active protect mode except that it impedes the functions of the ship, since the armor expands from a few inches thick to a few feet. Mouse told us to go to 'active protect' whenever we are under attack or at general quarters, and switch it off otherwise. We set up a demonstration with three ship panels—one un-armored, one in standard 'on' mode, and one in 'active protect' mode. The first two impacts from harpoon anti-ship missiles caused catastrophic damage to both panels and would undoubtedly have sunk or severely damaged a ship. The third panel, the one fully protected, had minimal damage and survived intact. The demonstration was very impressive."

The admiral elaborated on the adaptation of the invisibility mode to carrier operations, specifically the energy-absorption qualities of the radar shield nanolayer and the characteristics of the thermal-cloaking nanolayer. He was very excited about these effects. Invisibility in the radar and thermal-detection spectrums was more useful for a ship than mere physical invisibility. Since most air-to-ship and ship-to-ship weapons are now fired over the horizon, visual sighting of the target isn't typically required. Unable to target his fleet with radar or detection of the ships' heat signatures, the enemy would probably change tactics and use suicide runs at the fleet to release weapons using visual references. This was where physical invisibility would come into play, with the only visual clue being a wake trail in the water.

"We're going to need all of that protection, General, because we are going to be assaulted from every direction by hundreds, and perhaps thousands, of anti-ship weapons of various sizes. We'll face the combined forces of over fourteen nations. These forces won't be highly trained or well-coordinated, but they will be prepared to risk everything to find and kill us. I expect them to try many tactics: suicide attempts at ship-to-ship collision, augmentation with suicide speed boats, suicide runs with armed attack aircraft. Those are just a few of my nightmares.

"Even without the protection that your brilliant technology provides us, our naval forces could handle this threat with reasonable losses because of our stand-off capability and the new stealthy sea jets you have provided."

"What worries you the most, Admiral?"

"As we approach the entrance to the Red Sea and enter those narrow straits with our substantial naval force inevitably packed tightly together, we'll be surrounded by what amounts to twenty-five *land-based aircraft carriers*, because of the proximity of so many air fields in the Red Sea area. That's a real danger, unless we can neutralize it.

"Our Air Force assets at Diego will lessen the threat by attacking their land bases and their combined air forces with cruise missiles and smart bombs long before we enter the Red Sea. As the fleet gets into range, we'll launch additional sorties from the carriers to engage and destroy aircraft and bases. We're not likely to achieve total destruction of enemy assets, but we'll do great harm. Of course, they will rebuild airfields as fast as we cripple them, and hold assets in reserve until our forces get within range.

"It will be in the Red Sea that your technology plays a decisive role. When we enter the sea, the enemy will unleash every air asset available. The fleet will be bait for a decisive air confrontation, but our own air assets—coated with invisibility, and equipped with anti-radar and thermal cloaking—will have a considerable advantage. After that battle, either our fleet will lie at the bottom of the Red Sea or the enemy air forces will be destroyed."

"Will we win the sea battle?"

"I wouldn't want to be those poor bastards, General. They'll be slaughtered. We might lose five to ten ships, total, and perhaps one carrier, although a carrier is a tough beast to sink. The combined naval and air forces of the Islamic bloc will be destroyed in one decisive battle. We will prevail."

"Well done, Adm. Franklin. What else is happening in the world? Adm. Wainwright?"

"The world around us is definitely going to hell in a hand basket, General," Adm. Wainwright began. "It appears that a number of nations are trying to take advantage of the current situation to leverage our commitment to battle here. There are two main trouble areas: Israel and the rest of the world."

The admiral explained that Israel was being surrounded by an Arab army of more than one million and would likely be attacked at any moment—surely no later than the start of Scofield's foray into the Red Sea. The imminent Israeli war was directly linked to Scofield's attack. The Arab nations would destroy Israel once and for all during this cataclysmic clash of three cultures: Christian, Jewish, and Muslim.

The other conflicts were unrelated to the Islamic war, but had developed as one country or another sought to capitalize on the coming battle to press some regional advantage. Wainwright briefed the commanders on the developing conflict on the Turkey-Iraq border, where the Kurds were trying to assert hegemony over their tribal areas as Turkish and Iraqi forces moved towards Mecca. That area might explode into outright war soon.

China, still smarting from its defeat in the Indian Ocean, was moving naval and air assets into a posture highly threatening to Taiwan—whether to launch an invasion of the island or simply to draw US assets away from the Red Sea, was unclear.

Russia had sent troops into Georgia and Ukraine, and fighting had erupted in both countries. Whether these forays were mere feints, or the start of something larger, was also unclear.

Wainwright concluded his world tour with India's mobilization of substantial ground forces toward the disputed Kashmir territories. This threat looked real. It was highly possible that India would attack Kashmir when Pakistan, with the world's eight-largest standing military, deployed a large contingent of that army to Mecca. If Scofield was successful in neutralizing the Pakistani nukes, this conflict was sure to materialize.

Wainwright next described the heightened number and severity of terrorist events around the world, concluding: "The biggest uncertainty in the immediate future is what will happen to our area of operations when the attack against Israel begins."

"What do the Israelis say?"

"They're sounding a bit apocalyptic. I think they will fight a conventional war until they can no longer sustain it; then they will burn the Arab world around them with their nuclear arsenal. They're not happy with us for starting this mess, but it's unlikely that they will launch against us."

"That's quite a wild card, Admiral. What do you recommend?"

"I don't know, General, I honestly don't know. Maybe you should stroll over to Israel and talk with them."

The general paused to contemplate Wainwright's suggestion, then said, "for now, let's arrange a talk with the Prime Minister over a secure link as soon as possible, and let's continue our contingency planning for supporting the Israeli homeland defense. Let's hope our Islamic adversaries wait until we enter the Red Sea. If they do, we'll be in a much better position to aid the Israelis and prevent nuclear holocaust. Furthermore, I want scenarios for likely targets for the hundred-plus nuclear warheads the Israelis have. I also want specific projections as to how the world might react to such an attack, and recommendations for our response up to and including global thermonuclear war."

DOWN TIME

"Well, Mark, what did you think about the briefing?" Maria asked. They were alone in the general's private quarters.

"Jesus, Maria! What have we done? This whole thing is about to blow up in our faces. If the Israelis don't start tossing nukes, the Pakistanis certainly will. One side has to lose, and that side will launch a nuclear Armageddon."

"That's not my main worry. I wonder where the last two Iranian bombs are."

"That's right—they made four nukes, not two."

"Yep."

"And you don't know where they are?"

"Nope."

"That *is* a problem. The Iranians are crazy enough to use them preemptively."

"Yep."

"Even so, the Pakistanis and Israelis have many more nukes. Why isn't *that* a worry?" Mark asked.

"We intend to steal the Pakistani nukes, and the Israelis are not mad or helpless; we can contain both of those threats—unless, of course, we can't." In truth, the general could avert the certainty of an Israeli-initiated apocalypse only by neutralizing their nukes—a dilemma that would serve as the subject of future discussion.

"Is it worth it? The risk, I mean." Mark asked.

"What choice do we have? We cannot allow another nuclear attack on American soil. The two bombs in DC and New York caused utter chaos. I cannot think of anything more necessary."

"Hello, Mark," said the general as he entered. "Did you have a nice trip?"

"Yes, sir."

"Would you like to have lunch with Maria and me?"

"I'd be honored."

The base commander had relinquished his quarters to the general while he was at the base, and Mouse's minions had erected a nano-armored enclosure around the house. From this enclosure, Scofield could conduct the war and be protected against virtually any attack short of a direct nuclear strike.

Diego Garcia was on full war alert. An attack against the facility was both logical and expected, so Scofield's warriors wore their armored suits, as did the general, Col. Olsen, and Mark Osborne.

"Folks, while we're in the enclosure, we can safely remove our suits and get comfortable," the general said, peeling his off. The suit came off automatically. First, the helmet unzipped from the neck of the suit. Then, as soon as Scofield removed the helmet from his head, the suit unzipped itself—down one track on his chest, splitting into two tracks down the legs to each foot. He simply stepped out of the unzipped suit.

Mark wasn't prepared for what he saw. As he had expected, the general's body was finely muscled, the striations between his muscles extraordinarily well defined. His body was beautifully proportioned and symmetric, strikingly athletic. What struck the onlooker most, however, were the scars. They tattooed his body randomly: angry cuts, gouges, and punctures. Mark forced himself to look away out of politeness.

Maria was as beautiful as he remembered. Though she too had been wounded and awarded four purple hearts, she had no scars. He studied the contrast in these two perfect bodies, coveting both.

While Maria and the general casually donned robes, Mark shyly unzipped his suit, keenly aware that his body could not compete with theirs. He quickly slipped on his robe.

"Now we can eat in comfort," the general said. "These suits are so comfortable, and *comforting*, that Templar Knights sometimes don't want to take them off and become 'mere humans' again."

"I can understand that, General," Mark said. "The suit bathes you in comfort, makes you feel invulnerable, and heightens your senses to an extraordinary degree. And your skin is so wonderfully pampered in the suit."

"That's why we made the organic food fed to us though the suit bland. We figured we needed to give ourselves a reason to uncloak." The general eyed the feast laid out before them. There were definite perks to being the top dog.

Silently, the general and Maria attacked their food. Mark had never seen two people eat so much, so fast. They ate ravenously for twenty minutes, pausing only to smile at each other before launching another ten-minute assault. Mark had a sandwich, then sat back to watch the culinary orgy.

"Hungry, Maria?" Mark asked.

Maria looked up from her feast, her mouth full, and smiled sheepishly. She took a moment to swallow, smiling again, "Sorry, Mark. As I said, we Templars have ravenous appetites. We don't gain weight. So, when there's food around and we're hungry, you'd better get out of the way."

"I don't remember you eating like that on the ship."

"On the ship, we eat around the clock and never get too hungry. But the general and I've just traveled fifteen hours with only snacks to tide us over. This is a big logistics problem for the division. We're eating the Navy out of house and home, and the admiral is none too happy about it. He can't wait to get us ashore to stop the hemorrhage of his food stores."

The general, still eating, grunted affirmation. Finally, he sat back from the table and smiled. Loosening his robe to flaunt his distended belly, he belched and farted simultaneously, laughing heartily.

"Once a Marine, always a Marine," Maria said.

"Well, Mark," said the general, "you must have more questions by now. Fire away!"

"Okay. Now that I've experienced the nano-armor, I do have questions. Tell me about the computers in the suit. It seems to anticipate everything I do, or want to do. I think it can even read my mind."

"In very simple terms, it *can* read you mind."

The general expounded on the basic lattice nanobot building blocks that formed the foundation for the suit and other nanostructures: "The lattice nano-bot consists of three main elements: the core processing element, articulated arms, and claws. The core houses the tiny memory and processing elements and terminates the articulated arms. Each core has six articulated arm attach-ments for bonding to other lattice nanobots in three dimensions: up/down, forward/back, left/right. These arms contain communication filaments, which connect every two adjacent lattices to form a vast high-speed communication web that's virtually indestructible."

"How is it indestructible, General?" Mark interrupted.

"Do the math, Mark. Think about a screen door and consider how many ways you could trace a single cross-stitch of the screen—imagine that each cross-stitch is a lattice nanobot—from one end of the screen to the other. Many thousands of paths right?"

"Yeah," Mark agreed.

"Now imagine the screen door interconnected in not two, but in three di-mensions. The communications pathways are hard to kill." Scofield paused to see if Mark understood then continued. "Each lattice nanobot has ten free ar-ticulated arms to hold other nano-elements in place, move nano 'freight' from one lattice nanobot to another, or to construct bypass communication paths. The communication filaments are low-power electrical conduits that can push electrical energy around the suit, along with data. At the end of each articulated arm is a claw-like device that can grasp the end of another articulated arm (to bind two elements) or to bind or grab and transport nanofreight. With connecting arms, a lattice layer can be constructed to allow passage of material, claw over claw, throughout the suit. The suit becomes self-organizing, self-cleaning, self-repairing: effectively autonomous."

"General, how did you make an articulated arm so small and yet so capable?"

"The arm is a true work of engineering art. It consists of independent filaments woven together for strength and flexible at the joints. Each joint moves in a sphere of three hundred and sixty degrees in three dimensions. The filaments provide the core structure of the arm, with eight nanomuscle fibers attached at each joint. Like organic muscle, they expand or contract in response to an electronic pulse." Scofield paused again as Mark furrowed his brow.

"When I fell on my knee getting on the chopper to come here, the knee pad stiffened automatically and cushioned my fall. How can it be rigid sometimes and fluid other times?" Mark asked.

"The lattice muscle fibers can 'freeze' in a hardened position to provide semi-rigid isometric holding capability. This is sometimes used between lattice layers, as with the armor layer. Some lattice layers are loosely coupled to flex when stress is applied, as with the layer that massages the skin and removes waste. The six adjacent arms are single-jointed, and the ten free arms are double-jointed. When viewed through an electron microscope, the lattice nanobot looks like a spore with many tentacles. When you fell, the exterior armored layer stiffened while the skin protection layer softened."

"If there are so many independent computers in the suit, how does it manage them all?" Mark asked.

"Abstraction and hierarchy are the simple answers. Remember that each lattice nanobot has a lot of memory for a tiny gadget, and a functioning central processing unit or CPU, so each nanobot is somewhat autonomous computationally by itself. However, by itself, it could not perform a lot of complex processing; it's too small, too isolated, its processing set too primitive, and the overall system is too memory-restricted to do much by itself algorithmically. That's where abstraction and hierarchy are critical."

The general explained that the principle of hierarchy was used to assign to each nanobot a basic nano-autonomous processing program (BNAP). The

BNAP was like the basic input/output system (BIOS) in a personal computer that ran the lowest levels of the operating systems: the program that loads before Microsoft Windows or Linux. The BNAP provided basic building blocks to higher-level programs that wanted to use its memory, processing power, and communications paths or operate its internal or external functions, like the articulated arms. BNAP ran all the time, performing basic functions even without information from higher layers of software. It had important autonomous functions to perform, as well.

Scofield explained, "For instance, if a nanobot is severed from its neighbors, and from its supervisory programs, it will autonomously seek association with other nanobots, until it destroys itself or runs out of energy.

"On top of the BNAP program, a supervisory program layer is formed, called the lattice supervisor (LS). LS provides higher-level building blocks that an individual nanobot with its BNAP program could not implement on its own. The LS usurps the processing power from many nanobots, using the services provided by the nanobot BNAP layer to create a processing 'node.' An LS processing node acts like its own computer even though it's physically run on many smaller computers—the BNAP-driven nanobots. The LS program is redundantly stored; if the suit is damaged, it can restart itself without loss of stateful information. A suit has many different LS programs running simultaneously, each working on different problems."

Using the cloaking capability of the suit to demonstrate his point, Scofield described the LS layer function that sets up communication nets throughout the suit: "Cloaking requires that visual pixilated information from the nano-cameras in the back of a suit be sent to the front display nanobots to project the image to a forward observer. One of the LS layers coordinates this process. We've layered on top of that the nano-operating system, the OS," Scofield continued. "The OS provides us with higher-level functions. The LS and OS layers demonstrate the properties of hierarchy and abstraction very well." The general looked at Osborne, disappointed by his obvious confusion. "Well, you don't have to get all of it. Mouse can explain it better than I can. Do you want to hear about the global operating system (GOS) abstraction?"

Mark shook his head no. The general smiled, knowing that Mark would show keen interest in the GOS once he had observed its command-and-control capabilities on the battlefield.

"Suffice it to say that the software in the suit is the most advanced distributed operating system ever built. It has almost unlimited processing power. We've never used more than fifteen percent of it."

"Who programmed it?"

"Mouse."

"Mouse?"

"Mouse."

MODIFIED HUMAN

Capt. Deissler was looking at Bokert like a priest looking at a condemned man.

"We're about to inject you with a modified retrovirus that will forcefully invade each and every cell in your body over the next four hours," Nurse Furlong explained. "The DNA in every cell will be upgraded to a new version that will include the changes we've talked about. The moment that I inject your body with the first syringe, you'll experience a burning sensation that will be excruciatingly painful. This burning sensation will quickly spread throughout your body and become intolerable."

"By the end of the first hour, you'll feel nostalgia for the straightforward pain of the first injection," Nurse DuPont said. "At that point, every pore of your body will be screaming with pain. At the one-hour mark, we'll inject the second syringe and the pain will get worse."

"The good news," Capt. Deissler said, "is that your pain will reach a crescendo at about two and a half hours, then, after three hours, rapidly subside. The bad news is that you have about a ten percent chance of expiring during the first two and a half hours due to heart attack, stroke, infection, suicide, etc."

"Can't you put me out for this phase?"

"We could, but the process has evolved, and all Templar Knights agree that the process is cathartic, transformative for the soul, and the best and final test of a Recruit's worthiness. A person should experience this alone, without the support of others or the comfort of medicine. The torment will either kill him or make him a better and stronger Knight.

"Kneel, Templar Knight Recruit Gregory Adam Bokert," Deissler commanded. Bokert did as he was asked, frightened but resigned. "Receive the blessing of the nurses before your ordeal, accept the Lord God as your savior, and believe that He will protect you during this, your greatest test of will and strength."

"Bless this man who prostrates himself before you, my Lord, prepared to suffer greatly so that he may be worthy in your service," Nurse Furlong prayed with deep reverence over Bokert, placing her hand on his forehead as Bokert began to weep.

"O Lord, if his strength is insufficient, if you take him tonight, don't judge him harshly for his weakness. He is your child, a willing recruit deserving of your mercy, O Lord of Hosts," Nurse DuPont intoned, holding his head in her hands, gently weeping. "They are all my children, my Lord."

Nurse Furlong injected the long needle directly into Bokert's heart. He screamed as the retro virus pumped into his system and invaded his cells. Nurse DuPont placed an unsheathed dagger beside the bed and left with her two companions.

THE MOUSE RETURNS

"Hey, Mike!" Mouse smiled at the only mentor she had ever known. "I've missed you."

"You too, little one," Gen. Scofield said, flashing the smile reserved for Maria and Mouse alone.

"I hear your trip went well. I want a full brief later tonight on how you finalized the nanopaint to protect our navy."

"I'm ready now, General."

"Not yet. Let's talk first, catch up a bit. We haven't seen each other for almost a month." The general worried about Mouse on many levels.

"If you're wondering if I had any lapses, the answer is no. I was too busy to give in to my dark side." Mouse said this somewhat defensively, not hurt but disappointed; Mouse wanted their reunion to be happy—not morose or recriminating. The general sensed her disappointment.

"I missed you, Munchkin. I worry about you when you're gone. You're irreplaceable to me."

Mouse hugged the general tightly. She was in love with him and would do anything for him. She prolonged her hug, making Scofield uncomfortable at the implied intimacy.

"I prefer you not ask," Scofield said in reply.

Mouse was crushed. She wanted to give her mind and soul to Michael Scofield, whom she both feared and worshipped. But for his part he had always thought of her as a vulnerable little girl in need of his protection.

"I'm a virgin, Michael, but I'm twenty-two now, no longer a baby, and I'm ready to love someone the way you love Sarah and Maria. I don't have dark thoughts anymore. Your latest fix has put an end to those urges, and for that, I am grateful. We may not live through the next weeks, and I want to experience love."

"As well you should, Munchkin."

I'll always love you and only you, Michael Benjamin Scofield, Mary Louise thought hopelessly to herself. "Maybe Maria would take mercy on me," she blurted. "I'm not much to look at, but I'm very loving. I have a good heart once you get past the crust." *Boobie prize*, she thought, smiling at the pun.

Mouse was maybe four-foot-eleven, her face plain but not unattractive. *With a little makeup,* Scofield thought, *she could even be pretty.* Of course, as a holder of the keys to the genetic programming, she could make herself as beautiful as she liked, but she declined. She wasn't against genetic modifications, having accepted anti-aging and other non-cosmetic enhancements such as rapid healing and fast metabolism. She just steadfastly refused to change her physical appearance—except for the breast enhancement she'd bestowed upon herself two years earlier that she thought nobody noticed.

"You're most certainly ready, my dear. I wouldn't worry about anything. Just ask Maria. I'm sure she would be happy to help with your *problem*. Maria sees only a person's internal beauty, she is blind to physical appearance, and she loves you very much. I know this."

"Maria is a true friend, but I could not ask her. I'm too shy, and I could not live with the humiliation of her rejection. Would you ask her for me… please?"

Smirking, Scofield said, "Sure. I'll be your pimp."

"Now that *that* is out of the way, how are you, General?" Mouse worried about Mike Scofield's sanity as much as he worried about hers.

"I was visited by the Archangel Michael recently."

This news didn't faze Mouse in the least. "What message did he bring?"

"That I will die by the hand of the Beast or his brethren and that Maria will slay the Beast to avenge my death."

Mary Louise struggled to hold back tears. "So, it's happening." She said sadly, "Did he say when?"

"No. I assume during this coming battle, but I cannot be sure."

"Have the messages ever been wrong?"

"Often confusing, but never wrong."

"Did the Archangel reveal the name of the Beast?"

"Only that he was thirty years old yesterday."

"That's useful information." Mary Louise was a world-class data miner and sleuth. If anyone could narrow the list of possible candidates, she was the one. "I wonder why they're always so obtuse—these Angels in your visions. Why not just come out and say the Beast's name. That would make things easier."

THE NEW ALCHEMY

Greg Bokert longed for the pain to stop. Dazed, half out of his mind, he thrashed about on the bed, trying to cope, but it was no use. *Was it this bad for all Templars?* he wondered. Was he strong enough? He could hear his own pitiful cries as though they came from someone else. His eyes were drawn to the dagger, mercifully close on the bedside table.

Bokert's mind had clouded. The retro virus must have reached his brain and sown confusion. Hoping to slip into unconsciousness, but held back by the poison, he promised himself that he would persevere for another five minutes. Just five minutes more.

As he summoned his last ounce of willful defiance against the pain, the door opened and Nurse DuPont walked in. She mopped the sweat from his forehead and silently plunged the second syringe into his heart, emptying its contents. Greg wept uncontrollably as the contagion worked its way into his system.

Nurse DuPont stroked his hair and whispered a prayer that God might comfort and strengthen her patient for the trial to come. Crying soft tears, she left the room.

One thought emerged from the fog of Bokert's pain-addled mind: *Ninety more minutes of hell. God help me. Please help me.*

CHAPTER FOUR
THE WAR BEGINS

IN THE SITUATION ROOM

The general entered his makeshift command center to get an update on battle preparations. He found a buzz of activity. The two command center operators, Col. Olsen and Capt. Weaver, were glued to their headsets staring at screens displaying reams of data.

"The air and sea battles have begun, General," Maria reported. "The Air Force is prepping the airfields surrounding the Red Sea, and the Navy is probing the enemy forces guarding the entrance. So far, the battle has been one-sided, in our favor."

"Since it looks like we'll be here for a while longer," Capt. Weaver said, "I've sent for additional staff and communications gear to bolster our operational control capabilities. Col. Olsen and I can handle it until they arrive."

"One problem, General," Maria continued. "The Arabs are probing Israel's defenses. They may attack with full force at any minute. This would be a smart move; provoking the Israelis prematurely could disrupt our attack."

"They will do more than disrupt our attack if the Israelis start tossing nukes around," Scofield commented.

"Your flagship is now just two hours' flying time from Diego," Capt. Weaver confided. "It would be considerably safer on the ship than on this rock. Here, we're sitting ducks to a nuclear attack."

"We'll stay put for now, folks. I'm safe here as long as Maria is with me," Scofield said, cryptically. Both Maria and Weaver looked at him, pondering the meaning of his last comment.

"What's the status of my video meet with the Israeli Prime Minister?"

"Hopefully within the hour, sir," a control technician said. "She is somewhat preoccupied at the moment."

"Maria, how is the Air Force attacking the air bases?" Scofield asked.

"They're using stand-off weapons for now, cruise missiles and smart bombs that they can launch far enough out to avoid confrontation, since we've yet to establish air superiority over the Red Sea region. They're using data acquired via satellite surveillance and flyover reconnaissance to map the enemy's air defenses and to attack their infrastructure."

"Are the Air Force weenies being aggressive enough?"

"I'll keep a watch on them, General."

"You have my permission to thrash the commander, if necessary. I saw that the B-1s and B-2s were being shielded with nanoskins on the flight line. What'll be their operational profile?"

"Mouse has painted them with visual, radar, and thermal cloaking, and armoring against shrapnel or bullet-like impacts. The planes could not survive a direct missile hit, but if the missile can't find the bird, it can't kill it. She is also shielding the birds against radioactive, blast, and heat effects of a nuclear explosion. The birds can't get too close, of course, but they're better shielded than before. Mouse was able to add the extra layering without degrading the payload-carrying capacity of the birds by eliminating some of the on-board computers and batteries, and using the nanolattice to supplement and expand the computer systems and provide battery support. Mary Louise is a genius, General."

"Weaver, how is the Navy probing the enemy's naval and shore defenses?"

"I would describe their current effort as one of directed psychological warfare. They're sending in stealth sea jets to attack individual frigates and destroyers in a random pattern throughout their operational area. Two or three sea jets team up on a target and unleash a hellacious simultaneous attack. The target explodes spectacularly and rapidly sinks. The lamers haven't a clue as to what's happening, or how. They're panicking, and their losses are mounting."

"Good. What's the next step?"

"Adm. Franklin wants to talk with you about that. If we can continue to stand off and bleed enemy resources with minimal or no losses on our side, we should do that, he believes. We've yet to complete the Pakistani operation, and the Islamic forces are not moving to the battlefield as fast as we would have liked. By waiting, we can be better positioned tactically and destroy more ground forces if they're concentrated around Mecca before we land."

"Tell the Admiral that I concur. I want him to step up his interdiction campaign to include air assets, with direct attacks by the stealth aircraft we've prepared for him. I suggest he adapt the terror tactic of his sea-jet attacks to the air campaign. Coordinate with the Air Force on strategy and divide up the targets. Plenty to go around."

"Roger that, General."

"In the meantime, I'll be talking with Mouse about these new-fangled nanoskins for the ships and planes. Interrupt me as soon as the Israeli prime Minister is ready to talk. If she cannot make it to the video screen, I'll talk with my military counterpart or the foreign minister. Stress that it's imperative that we talk before they're too deeply engaged in warfare with the Arabs.

"Maria, I would like to have dinner with you, Mouse and Osbourne tonight. I'll continue to school Mark while you and Mouse catch up."

"Of course, General. What time?"

"Let's see how long the talk with the Israelis takes and if anything else comes up."

The general motioned Maria to step aside so he could speak with her privately, "Also, if it isn't too much trouble, Mary Louise would like you to initiate her in the art of love after dinner. She is ready to shed her virginity tonight. As she told me, she is no longer a baby. She is twenty-two…," he smiled.

"Oh, my God, I'm so honored that she would choose me. I'm surprised that she didn't want you to do the honors."

"She did, but that's a bridge too far for me," Scofield said.

"You bastard. After what she has done for you, you couldn't show her this kindness?" Maria's legendary temper flared immediately.

"I love her to death, but not in that way," Scofield said, a finger held to his lips.

"You can't see how she worships you? She hangs on every word you speak looking for approval and some sign of love. She has dedicated her life to you. That was fucking cruel Scofield." Maria was no longer whispering, to the enormous discomfort of everyone in the room.

"Thanks, Maria. See you tonight." Scofield lunged toward the door.

Maria glared at his back.

RETURNING FROM THE DEAD

Greg Bokert was dazed and drained, as if recovering from a terrible drunken binge. His body still burned, inside and out, and his nerves ignited if he dared move a muscle, so he remained perfectly still. *How long has it been?* he wondered. It must be after the two-and-a-half-hour mark; the pain was merely horrible now. He knew he had made it, he had survived. Soon the pain would be receding.

As each minute went by, Greg felt less pain. As he began to feel more human, he started to move his body and found that he was able to put stress on his joints again. Finally and fully awakened from his long nightmare, he sat up on the bed and saw Nurses Furlong and DuPont enter the room with Capt. Deissler.

"Welcome back, Templar Knight Greg Bokert," said Deissler. "You survived and are the better for it."

"God bless you, Greg Bokert," Nurse DuPont said. Nurse Furlong nodded and smiled.

"You'll soon feel better Greg," said Deissler, "and the changes your body undergoes will be dramatic. This is the payoff for your suffering. Usually, we allow our young Knights an extended rest, to repair their souls after this experience, but you must fly to Philadelphia today with Sen. Fitzpatrick. Sleep now and we'll waken you when it's time. Congratulations, Greg!"

NAVAL SKIN

The general walked the short distance to Mouse's laboratory and makeshift personal quarters: a cot near the corner, and a small refrigerator with a lamp on top. Scofield smiled as he thought about his mad-scientist protégé. Mary Louise was a child in so many ways, but she was also a driven engineer and scientist—and, when she was in her element, a force of nature. He had come to learn about the nanolayers Mouse had invented for the Navy ships and aviators.

"Hey, Mary Louise, have a minute for the old man?" Scofield had snuck up on her while she ranted on the phone with her engineering team.

Mouse jumped. She rattled off some orders and terminated her call. "Of course, Mike, I always have time for you, old guy."

"Smart-assed virgin nerd!"

"Old guy who sees dead people!"

Mary Louise looked for a hint about Maria and her special night.

"What? Oh, that! No, I forgot to ask. Sorry." Scofield paused but he made it obvious that he was teasing her. "Okay, okay, I asked. She said she was honored that you thought of her first, that she loves you very much and is looking forward to tonight."

"Oh, my God, Maria and me. I can't believe it. Thank you so much, Lion. How can I repay you?"

"Working harder, faster, and with less lip would be nice."

"If I worked any harder for you, I'd meet myself at the door coming in every morning."

"I don't have a problem with that... Okay, enough fun. I'm pressed for time. We may be interrupted at any moment for a meeting with the beautiful, sharp-tongued lioness who runs Israel. Tell me about your work with the Navy and Air Force."

Mouse explained in detail how she had constructed the nanoskins for the ships and airframes. Taking advantage of the programmability and flexibility of the lattice nanobot, she had custom-designed completely new autonomous systems for each ship and airframe. Her ship designs had been based on the vessels' original blueprints, but she soon realized that the blueprints were out of date, since the ships went through constant upgrades and refitting. When she had done the layers for Mike's amphibious fleet, she'd had plenty of time for her team to hand-apply the paint painstakingly, but time was too short on this project.

Mouse was clearly proud of the way she'd solved the problem. She'd pulled an all-nighter, programming an adaptive algorithm that would take its basic design parameters from the ship's blueprint and adapt them to the actual ship, as the paint was applied. The commander of her test ship went apoplectic when she commandeered his Navy frigate to wring the bugs out of her new algorithm. It took her fourteen hours to get it right, but that innovation allowed her team to paint over fifty ships simultaneously.

After she had the nanoskin primer painted on, it was easy to design specific nano-operational layers to the primer. She added stealth, radar-absorption, and heat-shielding layers easily enough. Next was the armor layer. She wanted the ships to be impervious to attack, so she designed a series of supporting armor layers to withstand a strike from multiple Exocet missiles and shrug it off. When she was sure she had succeeded, she went to work on her test ship.

It was then that she learned about ship-loading principals. As she was painting the frigate with her nano-armor layers, it started to yaw dangerously to port. The ship's captain descended on Mouse's makeshift laboratory, ready to go into orbit. She laughed at the memory of his hyperventilated tirade.

"Well, anyway, I learned in between his screaming that we were about to capsize and we would all be killed if I didn't stop painting his ship. As it turned out, I hadn't done the simple math of calculating the weight of trillions of nanobots."

"Oops!"

"Yeah, oops!"

Mary had to start over on the armor layers, and had to accept serious compromises. "That's why the ships are not protected against every threat: direct strikes from missiles can defeat the armor and cause serious damage. The admiral accepted the trade-offs. He recognized that the cloaking layers were a fundamental advantage and would save thousands of lives."

She had made one breakthrough, though, using an idea she'd stolen from Army tankers: reactive armor. She created a layer that would project three feet from the skin, when the skin was placed in active-protect mode. This layer would, in effect, prematurely detonate the incoming missile and cause much of its energy to be dissipated. The resulting strike would still cause damage, but with the interior armor and the ship's steel skin behind it, the damage would be lessened. The active-protect mode could not be left on at all times because it impeded ship operations, something that had provoked another fit from the frigate commander.

"Mike, if you give me six months, we'll stop building ships with steel. If I can save all that weight, then I can build an invincible ship!"

"Let's survive the next few weeks first."

"I'm constantly refining the programs that operate the nanoskins. I can update their software from here remotely. I think I can make the active-protect mode autonomous, so that it will trigger whenever an attack occurs without human interaction."

"Good work, Mouse. You did the impossible, under impossible time constraints. You are an American hero!"

Mary Louise had never received such glowing praise from the Lion before. She started to cry. She was embarrassed by her show of adolescent sentimentality but yearned for approval from her harsh taskmaster.

The general gently hugged his mad genius.

THE ISRAELI FACTOR

"Good afternoon, Madame Prime Minister. My name is—"

"I know your name, Gen. Scofield," the Israeli Prime Minister interrupted. "I know you have a birthmark on your left chest four inches from your nipple—"

"That birthmark was removed by a bullet that pieced my body armor in Afghanistan. You should have your files updated, Madame Prime Minister."

Prime Minister Ariella Abramsky more than lived up to her first name, which means "lioness of God." She seemed poised to burst through the screen and eviscerate the general. At forty-eight, she was strikingly beautiful, with the olive-skinned, raven-haired Israeli appearance that Scofield admired.

"What do you want, General? Isn't it enough that you have brought destruction upon my people? Will you not rest until you have caused Armageddon?"

"If Armageddon is near at hand, I'll not be the cause of it."

"How can we help each other?"

"I need the Israeli people to persevere long enough for my invasion to bleed our common enemies. Our greatest fear is that Israel will feel itself overrun and launch a full-scale nuclear attack, thereby goading the major powers into a global exchange of thermonuclear warheads."

"We'll defend ourselves with any and all means available. We won't lie down for any despot or group of despots who come to claim our lives or our freedom. The forces arrayed against us are greater than any we've faced in the past. We cannot hold our ground against such odds without nuclear defenses."

"You have over five hundred thousand active duty and reserve forces at your disposal, and as many more in your homeland defense militias. You have over thirty-five hundred tanks and over four hundred fighter and strike aircraft. You can win against the armies that confront you without initiating nuclear holocaust."

"I hope you are right, General. Time will tell. Nonetheless, I'll hold the nuclear option open at all times, until we are victorious or overrun."

"If you fire your nuclear weapons, where will they be targeted?"

"We'll launch directly at the invading forces and the population centers of the Islamic world. We'll set fire to Arabia and watch it burn from our bunkers, mourning our dead."

"If you wait until you are sure to be defeated, how will you attack the invading forces? They will be knocking at your door."

"That's the question. Do we launch when defeat seems imminent—or afterward, as an act of revenge? Frankly, I don't know the answer."

The general paused. He regretted that circumstances had ensnared this ally and empathized with their leader, whose people would be doomed if it came to nuclear war. "You must not act precipitously," he said at last. "Even if you are unsure if you can hold, you must act only as a last resort, to avenge your defeat."

"That's my decision, not yours, General."

"If you agree to my demand, I'll let you keep your nuclear weapons."

"You'll *let* me keep my nuclear weapons! Did I hear you right?"

"Capt. Buckhold, let yourself be seen by the Prime Minister," Scofield commanded.

A cloaked nanowarrior, at rigid attention, materialized next to the Prime Minister.

Shocked, and furious, at this unveiled threat to herself and her nation's sovereignty, the Prime Minister, a former Mossad agent, remained outwardly unmoved. She glared at the helmeted Marine. "Captain, stand over by the corner, out of my way," she ordered, "and be quiet."

She deserves to be saved, Scofield thought, impressed with her pluck. He gave Capt. Buckhold an almost imperceptible nod. Buckhold moved to the corner as ordered, and then cloaked himself again.

"Madame Prime Minister, I cannot allow Israel to run rogue here, or to burn the whole world because you *may* have no other hope. I'll take your weapons of mass destruction, if you force me to. However, I *will* defend your country. If Israel is burned to the ground, I'll incinerate the entire Islamic world, so help me God."

"How will you do that, General, should I agree to your plan?"

"Even now, we are probing the naval and air forces arrayed against us at the Red Sea and we'll destroy them before we invade. That may take some time. If the Arab nations aligned against you attack before then, you must defend your country by whatever means necessary—except nuclear or biological weapons. If you are attacked first, we'll immediately initiate our invasion, whatever the cost. This will draw their forces toward us and allow our southern conventional forces to close within striking range of Israel no later than seven days from the start of battle.

"Additionally, we have a carrier battle group with nanocloaking and armoring in the Mediterranean Sea near Israel right now. It's escorting two Army armored divisions embarked on the last amphibious ships in our arsenal. That's the level of commitment we are making to Israel. We're prepared to wed our forces with yours at sea, in the air, and on the ground, with our blood and our sacred honor."

"What of the nuclear weapons of Pakistan and the rogue weapons of Iran?" asked the Prime Minister.

"We'll neutralize the Pakistani nuclear weapons very soon. We don't know where the two Iranian nukes are located. We assume they are positioned in the United States to blackmail or punish us, or they are headed to this theater of operations. We'll visit the Iranian leaders and try to persuade them to give up their last two weapons. If they don't, and the weapons are used, then Iran will be destroyed."

Both rooms went silent as Ariella pondered the information presented to her. "I agree to your proposal, General. Please have your commanders coordinate with my command staff for the integration of your two armored divisions into our defensive line. How soon can we expect your forces to arrive?"

"The carrier battle group is in position now and has been conducting surveillance over your area of operations for two days. With your permission, we'll make the presence of the carrier known to our enemies and begin attacking their assets with prejudice."

"Let's let the commanders decide. We may not want to provoke our enemies until your divisions are in place and the activities of both forces coordinated."

"We can begin landing the divisions in two days."

"Are the two divisions equipped with these battle suits, General?"

"Sorry, no."

"We're finished then General, except for this: I can never forgive you for condemning tens of thousands of my people to death without giving us prior notice of your intentions. But I'll always honor you and your countrymen for coming to our aid in this, our darkest hour. I honor you for mingling your blood with ours on the battlefield."

"I ask your forgiveness, Ariella, and the forgiveness of your people–living, maimed, and dead. Whatever we might have done, it was destined to come to this. Better to face it now while the advantage is ours."

"Goodbye, General, and good luck."

DINNER AT EIGHT

Lunch seemed like eons ago to everyone in the general's private quarters. Each had participated in momentous events in the six hours since they had been together. The general and Mark looked tired, but Mary Louise was beaming. Maria, knowing why, smiled—a look not lost on Scofield.

Scofield glanced at Maria to see if she was still mad. If she was, she showed no signs of it.

"I suggest we get to work on this food before it runs away," the General began. "We can talk business after the feast."

Mark pushed backed from the table, to make way for the pigs. It took forty-five minutes for the three of them to clear the food, with help from Mark at the end. Scofield once again opened his robe, distending his stuffed belly, belching and farting as before. This time it was Mouse who said, "Once a Marine, always a Marine."

"Okay, Mark, let's you and I retire to my den and catch up. Maria and Mouse have work to do."

Maria grabbed Mary Louise by the hand and yanked her into Scofield's bedroom. Tossing Mouse on the bed, she glanced back at the two men and slowly closed the door.

"Well, I guess we're not welcome," said Scofield. "Hope they don't take too long. I need to sleep at some point."

"You can have my rack," Mark offered.

"No need. I'll push them to the side of the bed if I have to. I can sleep through almost anything if I'm tired enough—even felines mating. Meanwhile, we should have considerable time to cover a subject in depth. What would you like to talk about, Mark?"

"The nature of God and the universe and the role of Scofield's Templars in it."

"That's all?"

"Yep."

Mark had been talking with Templar Knights since lunch and was amazed that so many understood their mission under Scofield as a Christian one, more God than country. Their attitudes were so at odds with the American zeitgeist and modern secular politics that he found the conversations increasingly disturbing. It almost seemed as if Scofield were leading a cult, not an American combat division. All conversations had led back to the general. Scofield was the crux of this crusade. Mark used that word, *crusade*, even in his private thoughts. He pondered the spiritual commitment of these men and women to their cause, and their devotion to Scofield. The answer must lie with Scofield's beliefs. But would the general open up about them?

LADIES' NIGHT

"Thank you, Maria, for doing this."

"My pleasure, munchkin. Take off your hat, loosen that French braid, and comb out your hair. I want to see what you look like as a girl."

Mary Louise did as she was told. There were no further words as Mary began to unravel the massive French braid that reached to her lower back.

The beauty of Mary Louise's hair struck Maria. Thick, softly curled, dark auburn with streaks of red, it framed her face alluringly.

"You have beautiful hair, child. Why don't you wear it down more often?"

Mary Louise didn't answer this innocent question. Any revelation of the truth would divulge much more than she was able to at this fragile moment. She revered Maria and wanted her to think kindly of her.

Mary Louise approached Maria with her head bowed, saying nothing. She was so small next to Maria that she might have been kneeling. Maria caressed her hair, smoothing and combing it with her hands, luxuriating in its texture and weight.

Undressing her, Maria was surprised to find delicate lingerie beneath the boyish flannel shirt and jeans that had become Mary Louise's uniform. She gently removed the undergarments, saying nothing. Maria paused for a moment to look at Mary Louise through the prism of her deep friendship for her and the dawning realization that this fragile young women was worthy of her best effort.

WHAT IS GOD'S PLAN?

"Frankly, General, I'm disturbed by my talks with your men—the Templar Knights in particular," Mark began, "They seem more like members of a cult than hardened Marine Corps warriors. Did you alter their minds to be religious zealots when you gave them your genetic modifications?"

"Affecting a person's mental state or personality by genetic manipulation is difficult and fraught with risk," the general said. "Even if we could make someone more compliant, which we can't, we wouldn't do it. I've never served, nor will I ever serve, with anyone who wasn't first, and foremost, a volunteer. So, you need to find a different explanation for your observation. Do you have an alternative theory?"

"They believe in God, that you're His messenger, and that you have set them upon a mission assigned by the Almighty Himself. That's what they believe in a nutshell, but it doesn't explain why they believe it."

"What theory do you have to explain it?"

"I'm not a believer, General. So, I would have to say that you have deluded over ten thousand men and women into thinking that some deity wants them to murder a good chunk of the human race based upon your hallucinations."

Scofield's smile was thin-lipped. "Are you an atheist or an agnostic, Mark?"

"Atheist," Mark said. "I blame religious belief for most of the trouble in the world."

"You don't believe in any form of Higher Power?"

"No."

"No spiritual realm whatsoever?"

"I dismiss it as superstition, a support system for weak minds."

"I was an agnostic myself for a long time," the general said, "perhaps I still am to some degree. I was raised a Presbyterian, but could never get around my doubts; faith was never enough to overcome my rational mind. I blame my propensity for, and education in, the engineering disciplines for my lack of faith. I wanted proof, and God refused it.

"In truth, though, I've never understood atheists. Especially atheists who proselytize; they deny God, then judge harshly those who believe, like you're doing now. I understand well why a person might be an atheist, but I envy those who have faith."

"Perhaps weak-minded was unfair, but I do think it shows a lack of critical thinking. What would you say to a fourth-century BC Greek temple priest? Would you admire him for his belief that Zeus exists?" Mark asked.

"That the Greek gods are mythology doesn't mean that Christ didn't exist or that He wasn't God incarnate. Atheism as a functioning philosophy seems vacuous to me, offering no hope. While I've never fully embraced faith, I've always *hoped* that God existed. I think the need for God exists in all of us, and, because I cannot find an acceptable explanation for the perseverance of the 'myth' of heaven, I accept the mystery—if not the established explanations. However, I might be attracted to an orthodoxy that offered a unified theory, binding disparate religions together into a cohesive framework."

"You sound confused, General. If you don't believe, how is it that your men believe both in their God and in you as His messenger."

"A fair question. I struggle with doubt about the specifics of what God is but I've since reasoned that a God exists. I'll explain that in a minute. The correct term for my current belief system is that I am now a deist. So I've promoted myself from agnostic to deist due to those experiences you alluded to. Someday I hope to get a promotion to theist, which allows orthodoxy and is based on faith. Something is lacking in me that I don't understand..." Scofield's voice softened.

"That's convenient. Since you don't accept the doctrine of the church, your behavior lacks constraint. Is that how you justify having a mistress and murdering prisoners in cold blood while your men consider you some type of prophet?" Mark said. *I can play this game too Scofield,* he thought.

"Another fair question. Even as a deist I find much of what I do sinful. No, my struggle with faith isn't about gaining a license to sin. In truth, my love for Maria was birthed in battle, nurtured by an overpowering fealty that I cannot fully describe. Lust is the least of it. Deep intimacy born from shared pain, both physical and psychological, is closer to the truth. As for killing, I am a warrior, and warriors kill their enemy. I've never killed the innocent. In fact, I've sacrificed many of my men to protect non-combatants."

"Those three mujahedeen who served as your demonstration props were not murdered?" Mark asked acidly.

"You didn't see the eleven-year-old-Afghan girl they raped, then skinned alive. After they had finished with her, they raped her nine-year-old brother. I arrived on the scene and heard the cries of her brother and looked into the eyes of the little girl as she passed from this earth. In normal circumstances, I would have killed them on the spot. Something held me back that day; I took them and put them in a dark hole for years, waiting for the right time to release my rage."

"Holy Jesus, are you kidding me?"

"I smuggled her little sister out of the country. My wife and I raised her as our own. She is very special, but she will never escape that day or the place and circumstances of her birth. She, like me, is hostage to her dark memories. Unlike my other children—who have lived protected lives—she and I share a dark secret, never to be spoken of."

"How many memories like that do you have, General?"

"Many."

"How can any God allow such misery?" Mark asked.

"Another fair question. I've pondered this at times when I allow myself to remember such things. I don't have an answer suitable for you or myself. My greatest fear is that God is indifferent to our suffering."

"Or he doesn't exist, and this just happened for no particular reason except for a trio of sick fucks who happened by," Mark offered.

"They didn't just happen by, they tortured those children to lure my Lions into an ambush. Hearing her screams, we walked into the ambush willingly. One lion was killed, and another lost his leg." The general corrected. "I looked into that brutalized girl's eyes as she died. She went to heaven, an innocent child cupped in God's hands, a soul that had worth and mattered. Without that belief, I could not bear the memory."

"What happened to the enemy soldiers in the ambush?" Mark asked.

"To a man they died, save the three you saw that day."

"I can't decide whether you're a believer or a doubter..."

"We share that, then."

"Thank you for telling me the story, General. It helps to know that their punishment was just." Mark paused; his journalistic mind wanted to get the discussion back on track. It was clear to Mark that Scofield's diversion into this dark memory had changed the mood between them. "Is the Bible your authoritative reference for God's revelations?"

"I'm not bound by faith to the orthodox Christian church. Many faithful people believe the Bible is the literal word of God. Other faithful people believe that the bible is inspired text, but they allow for errors in the recording or translation of His messages through the ages. In the former case, that does present some problems for the faithful. I agree with the atheists on that point. There are many contradictions in the Bible, and it can be interpreted in many conflicting ways."

"Have you considered the atheist proof against God, General?"

"I'll try to repeat it. Correct me if I miss something: If God exists, then there exists a being that's omniscient, omnipotent, and perfectly good. If an omniscient, omnipotent, and perfectly good deity existed, then there would be no natural evil. But there is natural evil. Conclusion: God doesn't exist." Scofield paused.

"I believe atheists describe natural evil as things like earthquakes, landslides, sickness, and what not, as opposed to the moral evil that men do with their God-given free will—murder, theft, etc. Is that about it?"

"Yeah, that sums it up," Mark agreed.

"What is wrong with the argument?" Scofield challenged.

"It's a mathematical tautology. It can't be wrong," Mark said.

"Now who is being rigid? Who is to say that such a deity could not have reasons, perhaps unfathomable to us, for natural evil to exist?" Scofield didn't like intellectual rigidity. "So, why is the atheist 'proof' wrong?"

"Okay, I give up," Mark said peevishly. "You tell me."

"I can't blame an atheist for turning believers' own assertions against them. Orthodox Christians believe that God is omniscient, omnipotent, and perfectly good because the Bible tells them to believe it. There are those that believe that when Adam first tasted sin that he, and the millions who followed, corrupted the perfect creation of God. That may be the meaning behind the casting out of Adam and Eve from the Garden of Eden. His divine purpose may require natural evil, and perhaps Adam's sin was preordained. I'm not a biblical scholar, but that's not the point. In either case, God *may* exist, and the atheist proof against God becomes a bully's sophistry.

"In my view, the atheist who posited this 'proof' *is* a sophist. He won't point out to the faithful that believers may have misinterpreted the Bible or that natural evil may have a valid place in God's order. An agnostic will allow for this possibility because he has an open mind. An agnostic *wants* to believe, but he lacks faith; so he is free to ask tough questions, just like the atheist. Unlike the atheist, however, he doesn't suffer from the crippling effects of hatred. An atheist just wants to win the argument." Scofield postulated.

"So, you are asserting that the Bible isn't the infallible word of God," Mark said.

"I can assert that because I'm agnostic. To an agnostic, that conclusion seems obvious. The book, for me anyway, is too confusing and contradictory. The reasons are simple enough: through the years there have been too many translations between languages with imperfect semantic mappings, written and rewritten by men and man is, by nature, a flawed being.

"That doesn't mean that God isn't *reflected* in the pages of the Bible. Remember, according to my *deluded* followers, I've felt the presence of God, and I've spoken with angels. Therefore, unless I'm crazy or a liar, I'm an agnostic with a unique frame of reference. I'm left trying to explain the error in the atheist proof and the contradictions in the Bible because I *know* that God exists. The answer seems obvious to me."

"It's not obvious to me," Mark said.

"I believe angels have free will, like Man. I believe they've visited earth and spoken with men many times. I know firsthand that interpreting the meaning of their words is difficult. We should forgive the Bible's authors for any mistakes they may have made, since the angels are damn poor messengers. If they worked for me, I'd fire the lot of them.

"We see a reflection of God in the words of the prophets. But a reflection is, by definition, an imperfect view; His true nature is a mystery to us and may remain a mystery, perhaps for all time. Perhaps all he is asking is that we search for His true nature and in doing so we achieve some level of understanding and virtue.

"Meanwhile, the angels may be well-meaning screw-ups. They muck around, watching us poor humans try to cope with the physical universe and our moral selves. They try to help as best they can, but they're clumsy."

"This is all a convenient argument, General, but it justifies virtually any behavior since there are no divine rules in your world."

"Maybe so, but, if true, it would explain why there are so many expressions of God. Competing faiths may be nothing more than different views of the same phenomenon, from different viewers with different cultural biases. In any case, how does a lack of definitive rules differ from your atheist manifesto? If there's no God, there's no right or wrong."

"My atheist friends are usually more moral than the religious people I know," Mark asserted.

"Really? You must have special friends, then, because that's not my experience. But although you and your secular-humanist brethren espouse the model of rational behavior proposed by Ayn Rand, other atheists—Stalin and Mao Tse-tung come to mind—saw things differently and did unspeakable evil while demanding that the faithful reform their thinking to a more *modern* view," Scofield countered.

"No worse than the Inquisition. I think religion is all bullshit, General. I always have. I haven't seen anything to dissuade me since I've been with your forces. Nonetheless, I keep hearing from the troops that you have been visited by angels or by God himself. They believe that God has protected you from death on the battlefield—that you are touched by God somehow. If you are delusional—or, worse, lying to them—that's a great betrayal."

"If God is protecting me on the battlefield, I wish His shield were more effective. I have twenty-seven Purple Hearts and the memory of far too much pain. As for my 'visitations,' I could answer you, but it wouldn't matter."

"Why not let me decide, General?"

"You'll either believe me, or you won't. If you do, it will be because you want to, because you already have faith or want to have faith. If not, you'll rationalize my experiences as a near-death, low-oxygen-induced hallucination, reproducible in a laboratory."

"How *do* you know it's not an oxygen-deprived hallucination?"

"Some of the time, I fear that it *is*," the general said. Mark sounded conflicted to him, but not closed-minded. "Remember, I'm a scientist and engineer. I revere rational thought, just as you do, but I have an advantage. I'm a former nonbeliever who has touched the hand of the Lord our God. I have knowledge that you lack. I have the luxury of worrying about *why* God's plan is what it is, not *if* there is a plan. You're stuck on the 'if,' defending yourself against uncertainty by suspending curiosity and opposing any notion of God. As a scientist, I view your atheism as intellectual cowardice."

"Okay, General, I'm listening."

"Mark, you have seen many things in the past weeks that would have seemed magical, if you hadn't known there was technology behind it."

"You're a genius, but that doesn't prove God exists."

"I'm not a genius, Mark, but that's not the point. Lean over here and stretch out your arm."

Curious, Mark offered his right arm. Scofield grabbed it firmly. Mark could feel the grip from his powerful hands close on his wrist and his upper forearm.

The slight pain intensified and became excruciating. Mark tried to pull away, but the general's strength seemed preternatural. Looking Mark directly in the eye, the general increased the pressure on his forearm.

Mark screamed as he heard his radius bone snap, and then the ulna.

Scofield showed no emotion, no reaction to Mark's pleading.

"Oh, fuck, that hurts! Why did you do that? You evil son-of-a-bitch!"

"If I took your pain away, would that bring you closer to God?" The general's voice was soft, caring, and in control. He was totally focused on Mark, looking into his eyes while he supported his broken and twisted arm.

"If you can help me, please help me. It hurts. I can't believe you did that!"

Psychic shock can rewire a person's brain, for better or worse—Scofield had seen it many times in war. "Do you believe in miracles?"

"Fix it, or get me a doctor."

Scofield held Mark's arm in his hands and closed his eyes. He said nothing.

An unnatural heat permeated Mark's broken arm followed after a minute by the sudden relief of his pain. Mark's anger dissipated when his arm was healed, replacing his pain with awe. "Don't ever do that again," he pleaded.

"But now you understand that some things fall outside your rational understanding of the world. You're ready to listen. There's a mystery you want solved, and now it's deeply personal and compelling."

"Did God give you this gift?"

"I think he did, but I can't be sure. I didn't talk to God. I've talked to his angels, but God was a presence that I felt—we didn't 'speak' in any way I can fathom. When I left his presence, there was knowledge, but it wasn't clear how I had obtained that knowledge. At least, that was my personal experience. I can rationally explain my experience in God's presence as a possible hallucination when I was near death. What I cannot explain is how I survived such injuries, how I walked out of the hospital days later feeling better than before I was shot. I cannot explain why I've been able to heal the wounds of others from that day hence. I haven't found a rational explanation for any of that. Even an agnostic has to deal with the facts, and the facts known to me point to God. When I am being honest with myself I'm forced to admit that I know that God exists, I am thankful for that."

"So you do believe."

"I asked for proof, and he gave me proof. I graduated from agnostic to deist. I yearn now for true faith."

"If an all-seeing and all-powerful God *does* exist, I feel betrayed. How can he let the world suffer so much pain?"

"The only explanation I've come up with is that God may not concern Himself with the day-to-day affairs of Man or the physical world. God created the universe and started life, and within his plan was the emergence of Man; but he gave Man the greatest gifts of all—sentience and free will. He knew that Man would do terrible and wonderful things with those two gifts, but He must've been willing to pay the awful price for the horrors we've visited upon each other.

"Mark, we can talk about this again some other time. I apologize for what I did. Sometimes pain and shock can be cathartic, but I didn't have the right to make that decision for you. Please forgive me," the general said, helping Mark to his unsteady feet.

"I forgive you, General," Mark said, leaving hurriedly.

A RECKONING COMETH

The two women were still in bed, Mouse naked and sound asleep, snoring like a wounded bear. Maria smiled sleepily at Scofield.

"What did you do to Mark? I heard a snap, and screaming."

"I snapped his arm to make a point."

"Then you healed him?"

"Yes."

"Mark has his own path to follow. You should let him find it."

"I know."

"Go to sleep, General. The weight of the world rides on your broad shoulders, but you are only a man. I fear for your sanity, my love. These days of leisure will soon give way to unceasing action and danger. There will be little time for us. Mouse told me you have been visited again."

"Did she give you the details?"

"Yes."

Maria, stunning in her naked beauty, raised herself on her knees to look Scofield in the eyes. "Know that when you die, I die. I'll follow you to heaven to be with you. But before I join you, I'll eat the still-beating heart of the Beast as I watch him die mercilessly at my feet." Maria's eyes blazed as she spat out this devotion. She was fearless, and competent to carry out the threat.

"What makes you think I'll be in heaven? My soul has been darkened by war."

"Then I'll join you in hell, if that's where the Fates take us."

The two warriors gazed at each other, remembering the many times they faced death together, bled together, lost hope together. The bond between them was stronger than love. Tears flowed from the old man's eyes. Maria pulled his head to her bosom, and their tears mingled.

RESURRECTION

"**G**reg! Greg, wake up! It's time to go to work," said Nurse DuPont, shaking Bokert.

He felt like he was being pulled from a dark pit. He wanted to fall back into that restful pit, but she was having none of it.

"How do you feel?"

Greg took stock of his battered body and concluded that he felt absolutely wonderful. It was hard to describe. His joints moved without protest, his muscles felt strong and responsive, and his senses were sharpened in subtle ways he could not pinpoint. "I feel great, really great."

"The elixir is working. Each day will get better. I need to get you off to work. I'll be leaving you and Capt. Deissler here. Nurse Furlong has already returned to the battlefield, and I'll be joining her soon."

Greg peered inside his underwear at his seemingly unchanged member. "When does *it* regain it's, ah, vigor?" he asked with a smile.

Nurse DuPont remembered how it felt when she was going through the change, how wonderful it was to be so perfectly pain-free, athletic, and energetic. "It will happen gradually," she responded. "One day you'll wake up and you'll know. It comes as a pleasant surprise, bringing with it a renewed sexual potency. You'll be pleased with the results. I can tell you from personal experience."

"Of course," Bokert readily agreed.

"Okay, enough sentimentality. Get packed and ready to go. I'll make you some breakfast in the meantime. Your appetite will seem out of control. Don't fight it. You won't gain weight."

"Yes, ma'am. What rank are you anyway, Mary?"

"You can call me Col. DuPont, or ma'am, Private Bokert, but I prefer Nurse DuPont, or Mary."

At the entrance to the private jet facility, Sen. Fitzpatrick noticed the bounce in Bokert's step. "You look better than the last time I saw you, Greg. Feeling better, I hope?"

"Yes, sir. I must have lost five pounds retching over the last twenty-four hours, but my bug is gone and my hip is doing much better."

"Good for you! I need you at the top of your form for the next four days."

ALARMS

When Capt. Weaver entered the general's bedroom, he was amused to find Mouse naked on the bed snoring like a trucker.

"General, wake up. We have a situation." Weaver gently nudged Scofield, ready to leap back out of harm's way. The general had a nasty habit of attacking the messengers who woke him. Weaver wouldn't be the first the drowsy general had sent to the dispensary.

"Okay, Tim. Give me a minute. What's the ten-cent tour?"

"The probes against the Israelis are intensifying. We think the lamers are onto the fact that we're about to land two armored divisions to support the Israelis. They may launch a full-scale attack before we're ready."

"Damn! Okay, I'll be right there."

While waiting for Scofield, Weaver reached down to tickle Mouse's tiny feet.

"Hey, cut it out!" Mary Louise grumbled. Suddenly aware of her nakedness—and of Weaver's lascivious stare—she blushed from head to toe, pulling the covers up hastily.

"Mary Louise, you are a definite and certifiable munchkin hottie," Weaver said. "If you weren't a baby, I'd roll you over and show you how manly men do it. Obviously, Maria has no such qualms."

"I'm twenty-two years old! I'm no baby!" Mary Louise was clearly embarrassed; she wasn't bereft of shame like the rest in the room. All three looked at her and smiled, kindly but amused nonetheless.

"I'm sorry, Mouse. I was just kidding, I'm happy that you have been deflowered, and, frankly, I was a little shocked at how beautiful you are."

"I'm not beautiful, Tim, but thank you for lying."

"I would lie to make you feel good, but I'm not lying. You are very beautiful, and extremely sexy."

Mary Louise smiled involuntarily.

CHAPTER FIVE
WAR ON THREE FRONTS

CRISIS IN THE WEST BANK

"Show me the tactical map, Tim," Scofield said, leaning over the screen. Weaver brought up the current forces-distribution map and overlaid it with an intelligence map, to show both the known and presumed locations of friendly and enemy forces.

"What do the Israelis think?"

"They're worried, Mike. They anticipate a full-scale attack before we can get our ground forces in place. They have significant resources to thwart an attack, but they'd feel better with the Army divisions in place. Thankfully, the Arab forces are not ready. If they do attack, it won't be as coordinated as it should be. Against an organized and disciplined force like the Israelis, confusion could cause catastrophic losses. Maybe it would be best if they did attack. At least then, the Army can be placed where we know they're needed, rather than where the Israelis think they are needed."

"Maybe," Scofield said. "Maybe. But in that case, the Army will be offloading under attack. Can the Arabs reach the debarkation ports with their guns and missiles?"

"It's a tiny country, so you'd think the Arabs could reach any spot with their long-range artillery. In practice, though, that's not the case. Whenever a round is detected by the Israelis, their counter-battery radars immediately target the site for destruction. The Arabs were probing Israeli cities earlier, but they were suffering too many losses of important hardware and have backed off that strategy. We now think that the Arabs will wait to use their big guns until they launch an all-out offensive across the entire Israeli front. They can sustain fire once the counter-battery defenses are overwhelmed. The good news is that the Arabs are paralyzed and can't use their artillery or offensive missile forces to interdict Israeli defensive preparations."

"So, advantage to the Israelis. And we'll know when the attack is happening; because all hell will break lose at one time. Have the Israelis asked for support from our Navy?"

"Yes sir. They're using the carrier-based aircraft at night to camouflage our camouflage, so-to-speak, and we're having great success with our interdiction campaign to slow the Arab invasion preparations. The Arab ground forces are suffering grievous losses as they approach the front lines of the battlefield. We've stepped up that campaign to a point where we're starting to strain our resupply capabilities for aircraft maintenance and ordnance. We're working on that problem, before it gets pressing. We have to make sure that we have sufficient resources when the all-out battle begins."

"Can we get the counter-battery radar technology from the Israelis for our assault?"

"Sorry...no sir. The Israelis only have enough systems to protect themselves. As you know, after the Hamas and Hezbollah years of rocket attacks, the Israelis became experts at counter-battery warfare. I wish we'd listened and adopted their technology, because the United States isn't as developed in this area. We suffered from the 'not-invented-here' syndrome. We'll pay for that at Mecca."

"Roger that. It happened on my watch as the head of Science and Technology at the Pentagon." Scofield remembered the debate and wished he had interjected himself into it. "Thanks for the update, Tim. Now, tell me why you woke me from a sound sleep."

"The Pakistan operation is ready to go, and a decision to execute must be made immediately."

"What's happening that would change our timetable?"

"The Pakistanis are moving nuclear weapons out of storage and getting them ready to transport to the Israeli battlefield. Our intelligence has uncovered their current strategy. The so-called Islamic Defense Council (IDC), formed to counter the American threat and consisting of leaders of the nations sending troops to Mecca, requested that Pakistan release its nuclear arsenal to the council. The Pakistanis agreed, conditional on their retaining control until launch. The council leadership has ordered the deployment of nuclear weapons against the Israelis. They haven't decided to use them, but they want them ready, just in case."

"This intel from our spies at the IDC council?" Scofield asked?

"Yes. If they do launch, they intend to do it in complete surprise: their goal is one catastrophic attack to annihilate the Israelis before they can lash out with their own nuclear weapons. The strategy is motivated by two fears: while they worry about our sanity, they know there's nothing they can do to stop us from launching nuclear war against their cities. However, they are nearly apocalyptic about the Israeli intent to launch a first strike. They're convinced the Israelis will strike first, given the chance. Combine that with their determination to destroy the Nation of Israel once and for all, at any cost, and...well, it's easy to see what'll likely happen."

The general knew how the intelligence was being gathered and had no doubt of its authenticity. "Order immediate implementation of the Pakistan operation," he commanded. "Use any means necessary to secure all of the nukes."

"Yes sir."

"Prepare to move my headquarters back to the flagship as soon as practical. We'll follow the operation from our command center here, or on the ship in real time. If the op is still underway when we're airborne, we must have command and control capability in flight. Tell Franklin and Wainwright to increase their pressure dramatically on the Islamic fleet and be ready to up their timetable if needed. I want to close our position on the mouth of the Red Sea to within one day of striking the sea, in case of an Israeli attack or failure of the Pakistani operation. Order all Templar personnel to remain suited up while we're ashore."

"Roger that sir. Consider it done."

THE PAKISTAN OPERATION

Col. Melissa Franks, Scofield's chief of staff, had been assigned command and control over the Pakistan operation. She'd left with her team in two cloaked amphibious ships after the first staff briefing aboard the general's flagship. For three weeks, the team had infiltrated military bases, military and civilian offices, and command centers, tapping into the command and control and logistics arms of Pakistan's nuclear arsenal. Thirty-six hours earlier, she'd reported her conviction that the team had extracted every last possible piece of intelligence from the network and was in place and ready.

Just eighteen hours earlier, Melissa Franks had reported the news of the Muslim Council ordering the nukes deployed to the Israeli battlefront. Her Marines had become very effective spies in their cloaked nano-armored suits. They'd warmed to the task of eavesdropping and were taking wicked pleasure in recording the sexual peccadilloes of the leaders they surveilled.

Col. Frank's mission was threefold. Besides infiltrating the Pakistani military at all levels, she must be ready to neutralize it on command. She was also tasked with identifying the parties responsible for supplying the Iranians with the fissionable material used in the New York and Washington bombs, and to prepare punitive action against them. She wasn't daunted by the prospect of serving as their executioner.

To bolster her resolve, she'd stored a television news report of the explosions in DC. As the daughter of a Navy captain assigned to the Defense Communications Agency, she'd grown up in the Virginia suburbs. The violence done to her hometown, her family, and friends filled her with rage. It showed a once-beautiful, now mutilated young woman outside a hospital. The look on Marjorie Hampton's face entranced Melissa. She would have been about the same age as the daughter Melissa had lost in childbirth. Melissa had found her image and story enshrined on a Web site.

She would kill these evil men for Marjorie Hampton and her lost childhood friends. She would do so with great prejudice.

To accomplish her mission, she was allotted four Templar Division rifle companies from the First Battalion, Third Regiment—or *one-three*, as it was called—and two from two-three. These thousand men belonged to her husband's regiment; two companies shy of half of his regiment. Miles Franks had told her to take care of his boys and girls and to bring them and herself home safely. Miles knew his wife was capable. Except for Maria Olsen, there was no female Templar Knight more respected or feared than Melissa Franks. She was a tigress to Maria's lioness, and as deadly as any Templar alive.

Scofield knew that Melissa was the best choice for the Pakistan mission. It was critical she succeed and capture one hundred percent of the Pakistani nukes. He had told her that revenge against their leaders was important, but that it was subordinate to acquisition of the nukes.

"Are you ready, Melissa?" Scofield asked.

"I'm standing next to the Pakistani general in charge of all nuclear weapons. He has been working to move his weapons to Jordan, Syria, and Egypt in preparation for the Israeli campaign. He appears genuinely excited that he'll play a pivotal role in destroying the Zionist pigs."

She and Scofield were communicating over satellite encrypted communications links from the general's helmet to hers. Col. Franks was cloaked and her presence unknown to General Dharker. The communication feeds were multi-homed to the various command centers involved in the operation, so each could listen in.

"He'll soon regret his enthusiasm. Continue your brief, Melissa."

"The Pakistani order to mobilize the nukes has exposed sufficient intelligence to give us great confidence that we've located the eighty-three bombs in their arsenal. The General has most graciously shown us an inventory sheet indicating the exact location of all nukes. I'll send a copy of this document to your staffs now. I have teams standing by each nuke site...nine in all, ready to strike. The rest of the teams are located in the command centers where the nukes are controlled. Each nuke requires codes to set it off—highly secret codes accessible to only a few commanders. That's also a real advantage, since we can control the nukes by controlling their launch codes.

"The nukes that were on standing missiles have already been dismantled and made ready for shipment, and cannot be launched. If they've kept any in reserve, we'll find them, but they're useless without the codes to arm them.

"I also have teams at the President's and Prime Minister's residences, and at all regional governor's offices. The heads of the senate and national assembly are also surrounded. We've staked out the leadership in case we've missed any backup plans the government may have to defend their nuclear arsenal—and to execute the second part of my mission: justice for the murders of our people."

"Begin full, simultaneous assault on all targeted parties," Gen. Scofield commanded.

Melissa gave the command to attack. Each of the monitoring command headquarters spread through the fleet could hear the cacophony of orders and reports going up and down the line as Melissa's troops executed a simultaneous assault on the various targets. One by one, the reports came in of nuclear bombs captured, each report naming the specific bomb according to the inventory. Mouse's computers displayed a running tally against the published list as each report was assimilated.

As Melissa's teams captured commanders or political leaders, they would call out the individual's name and coded call signs to signal accomplishment of the sub-mission. External microphones on the suits of the Templar Knights picked up the sounds of shots as security personnel tried vainly to resist.

The main computers on the suits of the Templars decided locally what sounds to pass through and which to muffle, which helped the remote command centers manage the variety of inputs. Further processing at their end allowed Scofield and his commanders to make sense of the commotion. This was another of Mouse's brilliant computer programs in action.

God bless that little munchkin. Scofield thought.

Within five minutes, the Templar Knights had secured eighty percent of the nukes. After twenty minutes, all targeted leaders, codes, and command head-quarters—but one—were secure.

Melissa's team at Gen. Dharker's command center had not yet de-cloaked or grabbed their target. She held back, to see if Dharker had any tricks up his sleeve. Notified of the attacks at the four-minute mark, he had furiously called command headquarters throughout the Pakistani armed forces to mobilize for a counterattack. He didn't, however, try to activate a doomsday bomb or hid-den nuclear stockpile.

After ten more minutes, Melissa Franks was sure that she was in possession of the entire Pakistani nuclear arsenal. She waited to see what Dharker would do.

Dharker was despondent. He sat back in his chair, drew his pistol, and put the muzzle in his mouth. Melissa closed the distance between them in a flash. He pulled the trigger, but the hammer struck Melissa's thumb, not the firing pin. She yanked the gun out of his hand, grabbed him by the throat, and lifted him in the air like a rag doll. Dharker screamed in fear as Melissa decloaked and said in perfect Punjabi: "Not so fast, General—I have some questions to ask before I kill you."

The fact that a woman's voice spoke these words, and a woman's face stared at him, troubled the general as much as anything. He couldn't fathom how an Amazonian devil woman could pick him up off the ground and shake him like a small child. He was a brave man, but this experience was beyond understanding. He shook uncontrollably, bile rising in his throat, urine soak-ing his pants. He slumped down, resigned to his fate, releasing his body to her powerful grasp.

Melissa tossed the general aside. She then ordered all nuclear weapons per-manently disabled and booby trapped. This process had already begun, but a good commander always made sure. *Check, double check, then check again* was Melissa's motto. *No mistakes allowed today.*

With neutralization of all nukes confirmed, she reopened her link to Sco-field. "Part one of our mission is complete, General. I assume you have been monitoring."

"I have, Colonel. Tell your troops they did a fine job." The general was greatly relieved. This was one of many actions that had to go perfectly for his plan to succeed.

"Begin extraction of the nukes from Pakistan immediately. Navy is standing by with air and sea assets ready to assist. Destroy codes and command headquarters. Interrogate officers in charge of the arsenal to confirm that we have them all. Use any and all means necessary to ensure that they tell you everything. When you are done, kill them. After those orders are relayed to your strike teams and you are sure that they are executing the extraction to your satisfaction, arrange a parley with the political leadership. I want to be able to talk with each individually, and as a group or subgroup. Mouse will handle the technical details."

"Yes sir."

Melissa knew they could not hold these leaders long. Soon, the whole Pakistani army not deployed to Mecca would respond. Her lookouts reported significant activity at the nuclear bases, the command centers, and the homes of the hostage leadership. But she would do as she was told. "And a good death," she whispered to herself.

"Let's hope not, Melissa," Scofield said on a private face-to-face link.

"I'm sorry, Mike. I didn't mean for you to hear that."

"No worries. We'll play out this diversion only as long as we safely can, and then you can begin your egress. Adm. Wainwright, do you have everything you need to get the nukes out?" Scofield asked, bringing Wainwright into his conversation with Col. Franks.

"Extracting eighty-three nukes from nine locations in a country that covers three hundred and forty thousand square miles is a significant logistical operation, General," Wainwright responded. "We'll likely do it under fire, in any case, but if we wait much longer, we're certain to be attacked. I recommend that we move like our house is on fire."

"There are standing orders to destroy the nukes in place if they can't be extracted. Right Colonel?"

"Yes, General, but that would create quite a mess and leave fissionable material in Pakistani hands. I'll give these nukes up when they pry them from my cold, dead hands."

"I agree, Melissa, but make sure all booby trap devices are in place before they are moved. Full execute, Admiral, on your extraction plans. This is a priority for the fleet. Use any and all means to protect the extraction teams. And keep me informed."

THE INVASION BEGINS

"Nice job, General," said Adm. Franklin, beaming at Scofield from the plasma screen. "When Wainwright briefed me on what it would take to grab those nukes scattered all over kingdom come, I didn't see how you would pull it off. Your boys have made this invasion infinitely less dangerous." The admiral had run a number of scenarios for ending the conflict and extracting his forces, each one culminating in a nuclear confrontation with Pakistan. The snatch and grab of the nuclear arsenal had made all that planning unnecessary. *Thank God!* Franklin thought.

"I'm as happy as can be," Scofield responded. "I told Wainwright to give his exclusive attention to the extraction effort."

"I concur. My staff will coordinate with Wainwright to help in any way we can."

"Admiral, I want you to take whatever resources you are not using in Pakistan and apply them to the Red Sea push. We need to begin our final assault and invasion planning now."

"The fleet is ready, and we've upped our interdiction campaign. We can't say we own the seas or skies yet, given the significant reserves they're holding back, but whatever has floated or flown out to meet us is now at the bottom of the ocean. You have sucked them in, and we're killing them mercilessly."

Adm. Franklin was pleased with the operation thus far, but his enemy was beginning to counter his moves. They no longer came out boldly to confront unseen naval forces, but held back, perhaps waiting to strike the US fleet en masse in the narrow Red Sea waters.

"Given the revised timetable and the difficulties you face, when do we enter the Red Sea? And when will we land?"

"The best-case scenario is that we'll enter the Red Sea in three days, certainly under fire, although our forward elements will be entering within the next twenty-four hours. It will take another three days, at least, to let our naval strategy play out and finalize our domination of the air over the operational area. If we land your ground forces too early, the enemy will still have too many planes and bombs to attack them. By waiting, we restrict their targets to the armored ships, the anvil upon which their air wings will be broken."

"I thought as much. You heard my promise to the Israeli prime minister?"

"If we move the timetable up, we will still win. More of our sailors and Marines will die, but the victory will be as complete. I have a contingency plan for blasting our way to Jeddah as fast as we can steam there. That would be four days, versus six or eight."

"Good. Let's get on with it, then. I'll be leaving Diego to rejoin my flagship within two hours, just after my soirée with the Pakistanis. Keep me informed, Admiral. Tell your sailors they're doing a terrific job."

THE NEXT PRESIDENT OF THE UNITED STATES

The seven surviving senators were caucusing in their improvised chamber in Philadelphia. Once they had agreed on the rules and protocols for swearing in the seventeen newly appointed senators, Sen. Fitzpatrick began the swearing-in ceremony. The US Senate now had twenty-four senators, with seventy-six to go.

The immediate business was the passage of an act to allow governors to cast proxy votes for the new President. All had agreed that this dubious method (according to the Constitution, the House should've been electing the President, or an electoral college convened for the occasion) would be allowed just this once, and so certified in the first law passed by the twenty-four sitting senators (with prepackaged agreements from the governors).

With that important business out of the way, the body adjourned, to reconvene Wednesday morning for the presidential balloting. Senators and surrogates had one day to politick for the new President. Thinking the process was wrapped up, Fitzpatrick, Jones, and Tilden stayed in the chamber, jovially chatting with their peers. The evident confidence of Jones and Tilden made no impression on Fitzpatrick, whose political radar was turned low as he anticipated his inauguration. He believed that he had at least fifty-two solid commitments, and it was possible he would get many more votes as the momentum turned in his direction.

"Need anything, Senator?" Greg Bokert asked.

"Just make one last check on our committed voters, and let me know if we have any problems."

"I just completed a first run, and everything looks good," Bokert lied. "I'll screen one more time tonight."

"Okay, Greg. Call me only if you need me."

REVENGE

At their respective locations, two Templar Knights held each Pakistani leader in a painful vise grip. A third Templar Knight stood before him, providing a video feed to and from Col. Franks and Gen. Scofield.

"Today you'll receive the same quarter that my people received," Scofield began. "When I left the United States to come to your country, I pledged on my sacred honor that I would bring back the heads of the terrorist bastards impaled on the standards of my regiments. Your heads will be the first I take for my collection."

Realizing what fate awaited them, they began to plead for their lives—all but Gen. Dharker, who stared defiantly at his captors and the image of Scofield. Pakistan's president called out to Scofield to spare his life, claiming that he'd been duped by the evil men who'd stolen the uranium from his country.

Scofield signaled to Mouse to display the president's image to the other captives.

The president's pleading ended abruptly as a Templar Knight beheaded him. The Knight retrieved the head and placed it in a cloaked, nano-armored bag, where it disappeared from view.

As the others yelled in fear and rage, Gen. Dharker proclaimed above them, in perfect English: "We're not terrorists. We're politicians and soldiers of a sovereign country with rights under the Geneva Convention."

"You *are* terrorists," Scofield responded. "For too long, the Islamic elite have supported, directly and indirectly, the terrorists on your soil. Your failure to eradicate these vermin is collusion, at best, and makes you accomplices to their crimes. You gave enriched uranium and technical assistance to Iran. You knew they would use their bombs against us. This makes you complicit in the murders of over three million Americans."

"We're a sovereign country and may do as we wish with our natural resources. You are sworn to uphold the standards of the Geneva Convention."

"I rescinded our adherence to the protocols of the Geneva Convention before we left America. We seek revenge, not justice. We seek your destruction, not capitulation."

One by one, the Templars beheaded each captive according to his rank, leaving Dharker for last.

Dharker could not help but shake with fear when Col. Franks turned toward him. As she drew her sword and held it in the high ready position, she showed her face and said softly, "For Marjorie Hampton."

"Who?"

Dharker's face froze in a quizzical frown as the sword arced downward.

FAITH

"It's good to be back aboard ship. Don't you think, Mark?" The general was relaxing in an expensive silk robe, his white hair askew from showering.

"Yes, sir. I was getting too much sun in that damn tropical paradise."

"Before you say anything more, Mark, I want to apologize again for hurting you. I don't want you to walk around me on eggshells. I promise I won't do that again."

"General," Mark began slowly, "I've been reflecting on your message of faith. I want to become a Templar Knight and fight in God's army against our shared enemy."

"That's not the plan I have for you, Mark, but we have a rule in the Templars: we accept anyone who wants to serve and understands the sacrifices and commitment he must make. If you want in, you may join. I'm not allowed to say no to you, regardless of my objections. You can be a Templar if you can pass the severe rites of passage required. I warn you that at least ten percent fail—and when you fail, you die."

"I understand. I talked with Maria. She has explained the initiation rituals, and the excruciating pain of the genetic transition."

"And you still want to undergo this trial—to become a Templar Knight?"

"I know and accept the risks. Before we left Diego Garcia, I was baptized as a Roman Catholic, the religion of my parents that I rejected long ago to their great pain."

"Did you call them to give them the news?"

"They both lived in New York, General, in the Tribeca neighborhood in Manhattan."

"I didn't know that. I should've. I'm sorry. I'm surprised you hadn't mention that before."

"My parents and I have been estranged for over a decade. I buried my grief since the explosion; I haven't spoken about it to anyone. I think it has been a big part of my journey to you. I'd like to think they had a hand in it." Mark smiled.

"This is a big move for you, Mark. Are you sure you're not overreacting to the violence I inflicted on you last night or the news about your parents?"

"I asked for proof from God, and he gave it to me, with a brutality that I could not ignore. He sent me to you General, so that I would be saved." Mark smiled at the general, acknowledging the rhetorical theft of his demand of God. He had leap from atheist to theist, jumping over Scofield's doubt.

"How do you reconcile God's indifference to your suffering and the suffering of the world?" Scofield was probing for doubts he had never resolved in his own mind. "If you accept my indifference theory, it's unlikely that God sent you to me. He may not care about you in the least."

"Last night was a revelation for me, I know God exists and has a plan for me. I grew up a catholic; in fact I was quite dedicated to the church in my youth. I think the conflict I had with my parents and my church caused me to reject my faith. It all seems so sad now. All that time wasted. I cannot begin to tell you the joy I felt when I woke this morning, ready to embrace my faith again." Mark replied.

"I'll authorize an immediate induction for you. We must proceed tonight to make sure you are healed and ready for your role when we land at Jeddah."

"I want to fight by your side, General, when we land."

"You're not qualified to fight with me. No, I'll let you become a Templar, but it will remain a secret, except for a small group in my inner circle. You'll be of more use to us as a journalist and a writer, and to be useful, you must be seen as independent from us."

"I want to fight."

"You'll be trained to fight, but you'll remain a journalist. Remember the oath of obedience you will take."

THE BATTLE FOR ISRAEL

The First Armored Division, "Old Ironsides," had offloaded in the port city of Haifa, Israel, under the occasional bombardment of Arab aircraft and missiles from the north. The Syrians invaded southern Lebanon during the first stages of the Arab encirclement of Israel, linking up with the Hezbollah militias there and elements of the Lebanese army. Nearly a third of the Syrian army was at Israel's northern border, poised to strike, with another third in the Golan Heights on the northeast border. The final third (about 130,000 men and over a thousand tanks) of Syria's army had driven toward Jeddah to help fight the Americans there. All three Syrian army elements had been under constant harassment and interdiction throughout their prepositioning movements, with the southern Lebanon and Golan Heights army groups suffering the worst.

In accord with the strategy of the Islamic Defense Council, troops from bordering countries would predominate in the attack on Israel. Travel and logistics were easier for neighboring countries, and their motivation stronger. Pakistan, Iran, Saudi Arabia, and Iraq contributed the majority of their troops to the defense of the Holy Cities, while Syria, Jordan, Lebanon, and Egypt concentrated on Israel.

The First Armored Division was tasked with supporting the defense of northern Israel by staying in the reserve of the Israeli forces at the forward battlefields. Their job was to plug any holes in the Israeli defense and throw back invaders. Such a mission was difficult to plan for, and more difficult to execute. If called into the battle, they would be moving under heavy fire, into unknown dangers. It was a miserable job, and the First was well suited for it. They had the best equipment and best-trained armored crews in the world. They were ready, apprehensive, and determined. They also had an edge: American Marines in strange-looking uniforms had applied some kind of new armor plating to their tanks. The crews had one switch to control the new armor: cloaking on or cloaking off.

The First Armored Division was distributed along Highway Eighty-nine, which connected their operational area along a sweeping arc about three miles from the front lines. They would protect this corridor and the Israeli Northern Front at all costs.

With its armored and airmobile units, the First Cavalry Division (reinforced), the so-called "First Team," is the largest division of the United States Army. Reinforced with elements from additional armored brigades for the tank-heavy battlefield expected in Israel, the division landed at the port of Ashdod, Israel, south of Tel Aviv, about fifteen miles north of the Gaza Strip. The First Cavalry went inland to support the southern and eastern Israeli fronts in reserve, arraying themselves along Highway Six from just south of Ashdod to Netanya in the north, ready to block any advances from Egypt in the Sinai Peninsula to the south or from Jordan to the east. This position would also allow the First Cavalry to surround and defend Tel Aviv if things got ugly.

Both divisions were in place when the general conflict broke out.

IN THE BUNKER

Led by the nations of Syria, Egypt, Jordan, and Lebanon, the combined armies from ten Islamic nations attacked the state of Israel at 3:00 a.m. local Israeli time, exactly fourteen hours after the American capture of the Pakistani nuclear arsenal. For hours prior to the all-out attack, the pace of artillery and air attacks had quickened.

"Madame Prime Minister!" A personal aide shook Ariella from her sleep in her personal apartment, attached to the high command's bunker buried deep below Tel Aviv. Ariella glanced at the clock by her bed and realized that it was only 3:10 a.m.

"How long have I been asleep?"

"Just an hour, Prime Minister."

"What is it?"

"It started ten minutes ago. It's bad; the generals need you in the situation room."

"I'll be there in one minute."

Ariella didn't have far to walk; the situation room was just a few doors from her private quarters. By the time she arrived, her body had responded to the impending crisis with a flood of adrenalin. When she opened the door, she was alert and ready.

"General, what's the status of the battle?" she asked.

The general nodded at Ariella, looking tense and focused. "They opened fire on our front lines with over fourteen thousand artillery and tank guns. Our forces are being pounded. The enemy's used the bombardment as cover to maneuver their tank columns into attack positions."

"Nothing we didn't expect." Ariella was alarmed, but she knew, as their leader, that she needed to instill calm in these men. "And in the air?"

"We're being overwhelmed by the sheer number of enemy aircraft." Gen. Moses Weiss was a hardened veteran who wasn't easily roused or ruffled. For him to describe the battle in these terms was alarming more than the news itself.

"Our fighters will thin their ranks quickly." Ariella said hopefully.

"I hope so, Madame Prime Minister."

"I'm sure of it, General. Our men are better trained and better equipped."

"We have superior anti-tank weapons, better tank-to-tank weapons and targeting systems, and far superior aircraft, including US A-10 Warthog tank killers. But the sheer power and weight of the attack has caused our lines to break in many places simultaneously. Command anticipated the impact of the first hours of battle, and we planned defenses in depth, expecting to have our lines broken in some places. We hoped that the secondary lines would be called into battle later than the first hour but enemy forces have penetrated our front lines, so the secondary lines in many places have already been called into action. So far, the overall defense-in-depth strategy is holding, but we're concerned that concentrated attacks on the secondary lines could lead to a general breach." Gen. Weiss reported.

"Are the Americans engaged?"

"Not yet. We haven't committed the two American divisions to battle, but that's only a matter of time. If the secondary lines break, and the Americans cannot plug the hole, our only remaining defense will be the militias in every city and on every farm."

Ariella looked around at the frenetic activity in the situation room. Computers ran modern battles, Israel's battles originated in this room. Thousands—no, millions—of lives depended on the decisions that would be made here, right now. Ariella became hyper-vigilant, hyper-alert with each word uttered by Gen. Weiss.

"Okay… General, we'll overcome these initial setbacks and prevail. We'll prevail because we have no other choice. This battle will be won in three theaters: the north, the east, and the south. Give me the current status of each battlefield, one at a time, and we'll deal with each in turn."

"Yes. Madame Prime Minister. However, there is one other front, perhaps the most important front, and that's the air campaign."

"OK, let's start there."

"The battle will probably hinge on the outcome of the air contest. The IAF (Israeli Air Force) and American naval forces are engaged in a deadly, chaotic confrontation. Our anti-air missile batteries are somewhat paralyzed by the thick mix of friend and foe aloft. Unfortunately, the enemy forces aren't holding back, they're launching volley after volley of anti-air missiles into the skies above the battlefield, killing planes indiscriminately. Their willingness to sacrifice their own planes gives them a significant tactical advantage. Gen. Nadiv has suggested a solution for this problem and wants permission to proceed." Weiss motioned for Gen. David Nadiv to speak.

"Madame Prime Minister," Gen. Nadiv began, "at the present rate of loss, we could see the annihilation of the IAF in just a few hours. We must withdraw our forces west, away from the ground-based anti-air assets of the enemy, so that we can win the air battle through attrition with the inherent advantages we have over the enemy. If we do that, our anti-air command can fire indiscriminately at the enemy and inflict severe damage while our fighters shoot down those planes that venture into the Mediterranean to pursue us."

"Won't that leave our ground forces unprotected?"

"It will, but so will the annihilation of the IAF. Either way, they will be pressed by enemy aircraft in this battle."

"What about the American carrier planes?"

"They were being used for close air support because their cloaking allowed them to sneak past the enemy aircraft and missiles. We're now re-arming them for air-to-air combat to help us win the air war."

"How long will this repositioning be in effect?"

"Two or three hours—hopefully less."

"General, we may *lose* Israel in three hours!"

Ariella addressed the commander of ground forces. "Gen. Eilat, what do you have to say about this?"

"We cannot survive this level of onslaught without air superiority, let alone the complete absence of air assets. If we agree to this, your ground forces are doomed."

"And you, Gen. Weiss…where do you stand on this?"

"If the strategy succeeds, and we gain air superiority over the battlefield, then we can win this war. If we never gain air superiority, we'll most certainly lose this war and all of Israel. We've no choice but to follow Gen. Nadiv's suggestion. However, this strategy will be paid for in blood."

"How much blood, General?" Areilla asked.

"Many thousands of lives, Madame Prime Minister. Many thousands."

"Proceed with your strategy, Gen. Nadiv."

"This is madness!" Gen. Eilat stormed from the room. He returned moments later, seething but collected. Ariella nodded to him, forgiving his momentary lapse of composure.

"Gen. Weiss, how goes the battle to the north?"

"The enemy forces are penetrating along a wide front from the coast to the Sea of Galilee. The Golan Heights are under severe attack. There is general slaughter on both sides—more destruction suffered by the attackers than by us, but they can afford those losses; we cannot. I predict the complete collapse of the forward elements of the northern army in one to two hours. The American Division is positioned in the rear in support of our secondary line. It won't be necessary to send them into battle; the battle will come to them soon enough."

"Can the secondary line hold?"

"Maybe. I hope so. The Americans have significant armor assets in place, and they have their new technology to protect them, perhaps. We'll know in a few hours. Weiss continued. "The same situation exists on the eastern front with the Jordanians, Syrians, Iraqis, and Turks. They seem to have placed very large armies here to concentrate their attack. Our forces are reeling from the onslaught. They've already established two breaches in our line: one at Ma'oz Haim in the north and another at Jericho in the south. We believe the other attacks along the defensive line are feints and that they will pour their armies through these two openings.

Weiss paused to allow the prime minister to ask a question. When she didn't, he continued with his brief. "You'll recall our decision to defend Israel along our eastern flank at the Israeli-Jordanian border that extends between the Galilee in the north and the Dead Sea. We knew by incorporating the Palestinian West Bank territories within our lines that we would create an opportunity for a large insurgency action in the rear of our eastern lines. This fear has been realized. The entire West Bank is in flames; our troops are fighting two battles and losing both. I suggest we direct the entire American First Cavalry Division to meet these breaches head on and that we also release the West Bank Militias to the support of the eastern front and to defeat the West Bank insurgencies."

"Very well, General. I'll call the American generals and give them their attack orders. You may release the West Bank militias at your discretion."

The West Bank posed significant risk for the Israeli war planners. They could have established a defensive line to the west of it, withdrawing their civilian West Bank populations—more than 250,000 Israelis—into Israel proper and treating the West Bank as completely hostile, as they had done with the Gaza Strip. While this would have simplified the identification of friend versus foe, and kept supply lines short and in friendly territory, it would have made the eastern front untenable because of its sheer size. The West Bank extended deep into the heart of Israel, in spots barely seven miles from the Mediterranean. It would have been too easy for the Arabs to cut Israel in half.

The only viable decision was to defend Israel at the Jordanian border and deal with the insurgency issues in the West Bank—from Palestinians and infiltrated Arab fighters. The planners knew there would be attacks from two directions, but the overall defensive line would be smaller.

Ariella shifted her attention to her last battlefield. "Gen. Weiss, since we've directed the entire First Cavalry to the east, we won't be able to support the south with these nano-armored tanks. Can we hold the south?"

Southernmost Israel is a landlocked triangle anchored at the port city of Elat, extending northwest to the Gaza Strip on the Mediterranean and northeast to the southern tip of the Dead Sea on the Jordan border. Except for Elat, this area containing the Negev Desert was sparsely inhabited. There were three main roads—Highways Ten, Forty, and Ninety—all traveling north-south and terminating near Elat. Most Negev desert communities were scattered along these three routes. Highways Ten and Ninety hugged the Egyptian and Jordanian borders, while Forty ran up the middle. In the north were larger cities, and substantial population centers at Beer-shiva and Ashdod.

The Israelis had recognized that most of this area was indefensible. Their plan was to execute a classic scorch-and-burn organized retreat of the Negev to a defensive line extending from the northern tip of the Gaza Strip to the south of Beer-shiva and east to the Dead Sea. The positioning of the Beer-shiva line north of the Gaza Strip would put the Palestinians living there in a killing field.

The civilians who could not fight were evacuated from the lower Negev areas into Beer-shiva, and the rest of the militia units were organized into anti-tank killer teams, saboteurs, and sniper units. The regular army units dug a series of cascading defensive positions along Highway Forty to provide road bumps along the way for the invading Arabs.

"The enemy has attacked from the east and west of the Negev, with the Egyptians leading the attack from the west and the Jordanians, Iraqis, and Saudis from the east," Gen. Weiss continued. "They're penetrating along a wide front, but it will take hours to reach the Beer-shiva line. Our strategy is working well, we're already seeing supply line problems for our enemies, and the units tasked with delaying the advance are performing heroically."

"How are the fighters in the Negev getting out?" Ariella asked.

"Our fighters in the area south of the Beer-shiva line knew their fate when they accepted their mission. There is no escape for them. They will fight to the death defending their homes and synagogues."

"I'll call Gen. Scofield and the two American generals with the First Armored and First Cavalry. Keep me informed."

"Yes, Madame Prime Minister. Go with God."

"Go with God, General."

THE OUTCOME IS UNCLEAR

"Michael, we're at the crux of the battle. It will be decided in the next few hours." Ariella updated Scofield on the war that had been raging for hours along her entire border. "If we can hold the ground and maintain our supply lines to the front, we can start to turn the tide of war in our favor by counterattacking. Surely the Arabs cannot maintain the pace of their attack at these levels for long."

"Ariella, the Arabs have fully mobilized, and their forces are much larger and more capable than in previous Israeli wars. Most importantly, they're fully committed to battle. We have to be prepared for a prolonged attack, with their reserves put into action after your counterattack. They may even draw from their reserves surrounding Jeddah and Mecca. Without a significant victory in the air, we'll be in serious trouble. Your strategy to draw the planes out and to counterattack with the carrier forces must work." Scofield understood all too well the importance of air superiority. Throughout his long career, he had almost always enjoyed air supremacy. The one time he had not was a memory he preferred to forget.

"How long until the carrier battle group can re-arm and reenter the fight?" Scofield asked.

"At least an hour, General," one of the prime minister's staff answered for her.

"You are doing everything you can, Madame Prime Minister. For our part, we'll be entering the Red Sea with our forward elements within the hour. We're stepping up our timetable as fast as we can. Soon, we'll be on the ground, and all Arab eyes will turn to Mecca. We may be able to get additional naval air assets into your theater of operations in two days…maybe less. In the meantime, I'll redirect American Air Force bomber attacks immediately. The army units on the ground will coordinate the attacks with your forces. A few B-52 runs along the rear of the Arab lines should slow them up."

THE THREE ISRAELI FRONTS

"Gen. Weiss, I've talked to Gen. Scofield and the divisional commanders. What has happened since I left?" The battle had raged now for three hours; even in the bunker, the rumble of bombs could be felt. The computer screens placed high on the wall told a story of numerous breaches along the northern and eastern lines. In particular, the salients in the east had penetrated many miles into the West Bank Territories.

"We've thrown all of our reserves into the battle on both fronts. Either we stop them in the next few hours, or all will be lost. The northern front has been penetrated two miles deep along the entire Lebanese border. The Golan Heights has been completely overrun. We've ordered two brigades from the First Armored Division to the northeast to reinforce the secondary line along Highway Ninety, south of the Golan. They're under heavy air attack along their entire approach route and are sustaining significant casualties. The secondary line has not been breached as yet, but without the reinforcements, it won't hold. We're racing against time.

"We still have two brigades of the First Armored along the secondary line arrayed against the Lebanese northern front; they will be in contact with the enemy within the hour. We lost all contact with the forward units on the Lebanese border and the Golan Heights twenty minutes ago. They're presumed lost."

Areilla turned pale with the news. "*Lost?* How many soldiers are we talking about?"

"Tens of thousands, four divisions in total. There are bound to be pockets of resistance and survivors straggling back to our secondary lines, but initial reports tell us that the Arabs are killing all prisoners. We've also received reports of systematic massacres of the remaining civilians in these areas. I think they mean to annihilate us."

"Then they will burn in the hellfire of nuclear war," Ariella spat. "Inform all units of the massacres. Tell them there can be no retreat. We must hold the secondary defensive lines at any cost."

Gen. Weiss looked at Ariella, wondering if it would come to that. "We've already communicated the news of the massacres to the IDF. The response has been unanimous. To a man, the commanders have declared their intent to hold their ground or die in place. They will do their duty, as shall we." Gen. Weiss pounded the table, shaking with emotion.

"There is more, Madame Prime Minister, the situation on the eastern front is dire."

"Go on, General."

"The two breaches of our lines along the Jordanian border have developed into major salients. The enemy is driving towards Afula in the north and Jerusalem in the south. The insurgencies within the West Bank have hampered our abilities to counterattack. The Afula salient is the most threatening to us from a tactical perspective; if they decide to turn north, they can attack the rear of our lines that are desperately defending the Lebanese front and the Golan. If they do that, the northern forces will be cut off and annihilated."

"Do they show any signs of turning north, General?"

"Not yet, thank God. Their strategy is probably to cut through to the sea and link up with their northern forces in an attempt to cut us in two. They're probably unaware of how vulnerable we are in the north. We have the option of pulling the two American brigades that were sent to support the Golan secondary line to attack the northern flank of the Afula salient. We'll lose the northern front, but it could succeed in pushing back the Afula invaders."

"Where are the two brigades now?"

"They're about to engage the enemy along Highway Ninety. We should be getting reports in soon regarding their outcome."

"Where are the First Cavalry units?"

"They've split into two pincers to attack each salient head on. They've been ravaged by enemy air attacks but have moved close to the enemy with their forward units and have begun to engage. We should know soon who will prevail."

"What do you recommend, General?"

"Stay the course. Let the First Armored attack the Golan invaders and hope the First Cavalry and our militias can turn the tide in the West Bank. If we send the brigades south to Afula, we buy time but guarantee our eventual defeat. We need something to turn the tide in our favor; if we can gain air superiority, then we might have a chance."

"Gen. Nadiv, is our air strategy working?"

"Yes, Madame Prime Minister. We've ravaged the planes in the northern and eastern fronts and have refueled and rearmed the nano-armored American carrier jets. We'll be counterattacking within the hour."

"Move up your time schedule if possible, Gen. Nadiv. Everything hinges on your success."

"There will be no delays. The IAF commanders have monitored the reports from the front. All IAF forces will be ready; they will strike without fear and will show no mercy."

"The only good news is in the south," Gen. Weiss interjected.

"The Negev desert attrition campaign has worked better than we had hoped. The enemy units are strung out along the entire line of approach. We've been able to gain a semblance of air superiority in the south because the enemy's air-to-ground missile systems haven't been able to keep up with their advance. This has allowed us to interdict the enemy along his entire attack axis. The forward enemy units have reached the Beer-shiva defensive line and have been decimated. They've fallen back to regroup—and, presumably, to let the rest of their forces catch up."

Ariella rose from her seat to speak to the men and women in the command center. She paused while they turned to look at her. "I'll leave you while I go to authorize the preparation of our nuclear arsenal for our defense. Our nation will survive or perish based on what you do in the next few hours."

THE NUCLEAR OPTION

"What are the options, Gen. Sharett?" Ariella was speaking to the general in charge of all nuclear forces in Israel.

"We can exact wholesale retribution on the Arab world and burn our homeland so that no one shall live here for a hundred generations. We have enough bombs to burn the Islamic world into a molten slagheap. What we don't kill, we can cripple with EMP explosions in near space."

"Explain EMP to me again, it's been awhile since I was briefed?"

"Electromagnetic Pulse. We explode a bomb two hundred miles into space, and the energy from the bomb interacts with the lower atmosphere to create an electromagnetic pulse that wipes out all electricity and most electronic components. We've developed specialized nuclear weapons to maximize the destructive effect of EMP. Our weapons will fry the electrical grid, kill communications, and stop most cars and trucks in their tracks. The means of production are destroyed without killing one person. The mouths still need to be fed, but there are no means to produce or distribute food. The aftereffects of an EMP strike will devastate the Arabs. If you seek revenge, I have it ready for you."

"Can the Americans stop us?"

"They're not omnipotent, Madame Prime Minister. If they were going to stop us, I think they would have done that already. All reports are that we have complete control and security of our arsenal. The missiles and designated bombers are ready and waiting for your command."

"Name your primary targets."

"We've targeted all Islamic cities with a population of one million or more. We have eight EMP bombs ready that will fry every car, radio, and toaster in northern Africa, the Mideast, Turkey, Pakistan, Iran, and Indonesia. We've also targeted EMP bombs for India and China for good measure. We'll leave this world a fifteenth-century remnant with no means of production to feed the billions who have sought our destruction."

"What happens to Israel?"

"For those who don't die immediately, we can expect a full retaliatory attack after we fire our arsenal. No one will survive; no one will be alive to repopulate our homeland."

THE CRUCIBLE OF BATTLE

"Ariella, the enemy has reached the secondary line along Highway Eighty-nine, where our forces and the American First Armored are engaging them. We have reports coming in from the defenders confirming that the remnants of our forward forces are attacking the flanks of the enemy columns in small groups of tank hunter-killer teams. Some of the forward forces must have survived the initial attacks. Col. Allon radioed in and said the fragmented teams were throwing themselves at the enemy in suicidal attacks. They're being slaughtered, but they are having a dramatic effect on the enemy formations, forcing them to turn and defend while trying to attack. The colonel collapsed in tears after sending in his report."

Gen. Weiss was also emotionally overwrought. His son-in-law was in a unit on the front lines that had absorbed the initial shock of the Syrian, Lebanese, and Turkish attacks at the start of the war; there had been no word from his unit for three hours.

Just then, an assistant to Gen. Weiss appeared and excitedly relayed a message to him.

"What is it, General?" Ariella asked.

"The American nano-armored tanks have fully engaged the enemy on the northern front. The enemy tanks are blind to them and are being systematically annihilated. The American commander has ordered his tanks to attack. They've left the line and are moving to engage. It appears the attack along Highway Eighty-nine is faltering, and the enemy is breaking ranks."

A cheer went up in the room.

Gen. Weiss turned to his chief of staff. "Colonel, tell the American commander to return to his defensive position after he has repelled the attack. We'll need his unit intact later."

"Yes, sir."

"What about the Golan Heights, General?" Ariella asked, hoping for similar news.

"The two American brigades have arrived and—together with two of our reserve divisions entrenched along Highway Ninety—are preparing to engage. The enemy has not halted their attack after overrunning the Golan. They attacked the secondary line without pause and are attacking two positions with large armored columns as we speak. We still have no word from our Golan divisions."

"Gen. Nadiv, how is the air war shaping up?"

"Madam Prime Minister, it's with great pride that I can report that the IAF and the nano-armored American carrier-based planes have ripped into the remaining enemy air formations. We've killed at least one hundred aircraft, and their forces appear to be retreating into friendlier airspace. This has allowed our attack aircraft to resume support of the battles at Highways Eighty-nine and Ninety and along the two salients, Afula and Jerusalem. We'll maintain this level of operational tempo until we are victorious. Additionally, the American strategic bombers have been attacking the rear of the Jordanian lines to try to cripple the attacks into the salients. They're having great effect; it's a target-rich environment, as the Americans like to say."

"Do we own the air?"

"Not yet, but I'm confident of victory in the skies immediately over our country."

"What are your aircraft losses?"

"Significant. We cannot survive a second wave like the one we've just had. Let's hope they save the rest of their aircraft for the Americans. We have reports that the American fleet has probed the Red Sea and that aircraft are being diverted in that direction. If that trend continues, then we can prevail, even with the meager force that remains.

"The American nano-armor planes are magnificent; they've had only five aircraft losses, and they've inspired fear among the enemy pilots. They've turned the battle in our favor."

"Madame Prime Minister," began Gen. Eilat, commander of the ground forces, "I have reports from the lines at Highway Ninety saying that the first IAF ground attack jets have hit the enemy assaulting their lines. The whole front erupted in cheers simultaneously. The return of the IAF, something my boys have been praying and dying for, has boosted morale and fortified my men immeasurably. It couldn't have come at a better time."

Gen. Weiss broke in: "A report from Highway Ninety has come in. It appears that the American commander arrived with his lead elements and found the enemy column strung out in a long line, preparing to attack. He took his cloaked tanks and left behind his support units and thrust into the rear of the enemy formation. Our forces attacked simultaneously from the front, while the First Division tanks rolled up the rear. The attacking enemy force was annihilated, and its rear elements have retreated. The cut-off Muslim forces were not accorded the option of surrendering."

"I understand, General. Today we focus on survival."

"The south is holding well. The enemy is stalled south of the Beer-sheva line, unable to penetrate our defenses, and the shift in air superiority will likely make that stalemate permanent. Now we must hold the east and the West

Bank. To do that, we must repel the incursions. We should be getting reports from the front soon."

"If we don't hold those lines?" Ariella asked.

"Then they will cut us in half or by thirds. The gains we've made in the north and south will be reversed, because one or the other will be attacked from the front or the rear and will be cut off from supplies. If the enemy reaches the sea, we are doomed."

"Then they won't reach the sea," Ariella said defiantly. "Thank you, General. I'll contact Gen. Scofield and ask for more strategic bombing runs and faster penetration into the Red Sea to divert the enemy further. In the meantime, prepare a preliminary estimate of our casualties, military and civilian, and an estimate of our ability to withstand a second attack along the two fronts that have held."

"We have those estimates tallied in real time throughout the battle; I'll digest it and get back to you soon. Prepare yourself for bad news."

THE WEST BANK

"Madame Prime Minister, the counterattack against the two western salients will consist of three separate but simultaneous attacks. The First Cavalry split into two attack pincers and headed straight towards the salients to roll them up and repel them from Israel. They're in place now, ready to attack.

"Simultaneous with that attack, the surviving Israeli units on the Jordanian front will attack the rear of the salients to cut off reinforcements and supplies. This is likely a suicide mission for the eastern front units, since they are surrounded and maneuvering in intensely hostile territory.

"The third action will consist of a full-scale release of all militia fighters in the West Bank and along the flanks of the two salients. They will attack and harass the Arab forces from all angles while the two counterattacks are launched. We're ready to launch the attack."

"By all means, Gen. Weiss, attack!"

The activity level in the room became even more frenzied. Ariella was amazed at the noise and the energy level of her military staff, who had worked four straight days without rest. She heard Gen. Weiss tell Gen. Nadiv to concentrate at least ninety percent of his attack aircraft on the salients. Gen. Nadiv looked towards Ariella, who nodded her approval.

Ariella could see the progress of the battle on the tactical screens, but wasn't schooled in the detailed reading of the symbols representing the many diverse battle units. She knew that the symbols represented real people fighting a real war. Many thousands were dying gruesome deaths as she stared at the screen.

She had given the order to attack. Suddenly, the weeks of stress surfaced and began to overwhelm her. She had understood her role in this immense tragedy throughout the day, but, at this moment, she felt it on an emotional level that she could not will away. Her usual deep well of inner strength and self-confidence was no longer protecting her from an overwhelming feeling of alarm and fear. *What if we fail?* she thought, wanting to weep but knowing she could not.

One of the staff, noticing her apparent confusion, moved closer to her position and started to describe the action on the screen. "The blue squares represent friendly maneuver units. We can monitor the location of unit headquarters and subordinate maneuver elements through radios they are carrying that have embedded GPS location finders. This part of the operation is automatic: the commanders don't have to call in and tell us where they are. The red icons represent our best intelligence about where enemy units are located and where they are maneuvering. When they're blinking, as that red square is over there on the lower right of the screen, that means we have updated information that an operator needs to look at."

The colonel paused to see if she wanted to continue. Ariella nodded. "We have numerous sources of intelligence that has to be processed and integrated to the tactical map. Some of that's done by computers automatically and some by humans processing real-time reports from the field.

"It can be quite frustrating sometimes, because the human intelligence is often in error and late. The map can only approximate the tactical situation.

Still, it's better than what we had in the past. The fog of war has not been lifted by technology, not yet anyway, but it helps.

"We're the top of the food chain for intelligence, and our processed and coalesced information is sent back down to our subordinate units via encrypted links to provide the operational commands with the best information we have. If we make a mistake, people die."

"Why has that blue square turned opaque?" Ariella asked.

"There could be many reasons. That particular square represents a mechanized brigade of the One-hundred-and-sixty-second Armored Division from the Central Command. It's possible that we've just lost communications with the brigade or there is new intelligence that has come in but has not been updated. Would you like me to check on it, Madame Prime Minister?"

"Yes."

The staff colonel, a handsome and sturdy fellow, went to one of the operators and discussed the situation. He returned looking grim. "The One-hundred-sixty-second was in contact with the enemy attacking the Afula salient rear; the Four-hundred-first Brigade, which is the square represented on the map, was the lead maneuver element. We've lost communications with the brigade headquarters, and now with the divisional headquarters as well."

"Is that common? How long till communications are restored?" Ariella noted that the other blue squares in the One-hundred-and-sixty-second Armored Division sector had also turned opaque.

"It has been a common occurrence today."

"Meaning?"

"The unit has probably ceased to exist as an organized command."

"Do you know anybody in the One-hundred-and-sixty-second, Colonel?"

"My son is a tank commander with the Four-hundred-and-first Armored Brigade, and my brother-in-law commands the First Battalion of the Nine-hundred-and-thirty-third Infantry Brigade."

"I'm so sorry, Colonel."

"Thank you, Madame Prime Minister."

"Go back to your work, Colonel. I can figure things out from here."

GRIM PREDICTIONS

"**M**adame Prime Minister, we have the preliminary casualty reports and readiness status for the three defensive fronts." Gen. Weiss motioned to Ariella to join him in the soundproof briefing room just off the command center's central operating area.

"What's the bottom line, General?"

"It's impossible to count the actual casualties, but if we assume we've lost eighty percent or more of the men and women in the units that have been cut off, and sixty to seventy percent of the combat-ineffective units, then the dead will number over eighty thousand, with two to three times that in wounded. Our hospitals are overwhelmed, so many wounded will die. The majority of the casualties are from regular-force units that populated the front lines, so our readiness against another attack is low—very low in the Northern Command. As you know, the situation in the Central Command and the West Bank is critical, and we've yet to repel the invaders, let alone tally our ability to sustain another attack. The south looks good, and the overall air campaign has been won."

"Eighty thousand dead in eight hours...a quarter of a million casualties? How is that possible? How can so many die so fast?"

"Madame Prime Minister, we are actually lucky that it hasn't been worse. The loss of personnel is significant, but the material loss is even more critical. It will take years to resupply the equipment after this is over. If the enemy decides to attack the northern front again, in force, we don't have the ability to defend it. The Northern Command will be annihilated." Gen. Weiss looked weary and old, near exhaustion.

"What about the American First Armored Division?"

"They've suffered thirty percent casualties, and their supply lines are exhausted. They can be very effective defending the positions they currently hold, but will be ineffective if the tide of battle should go back to the same intensity level as before."

"And if we win in the West Bank?"

"If we beat back the enemy in the West Bank, *and* they decide not to counterattack in either location, then we might prevail—if the Americans can enter the Rea Sea in time. We can also hope that the Muslim armies are in worse shape then we are."

"Thank you, Gen. Weiss. You must win the West Bank."

"I know, Ariella. I know."

SLAUGHTER IN THE WEST BANK

66 The attacking armies in the Afula and Jerusalem salients are feeling the effects of the combined attacks from their front, flanks, and rear. It appears our efforts to cut off their supplies have been successful, and the lead attack columns have slowed to a stop. The American First Cavalry Division and their nano-armored tanks have enveloped the lead enemy units and are attacking as we speak." Gen. Weiss had turned from his command desk to brief the Prime Minister.

"The West Bank militias are fighting house to house and street to street with Palestinian and Arab infiltrators and yet, somehow, have also managed to reach the flanks of the regular forces. They're fighting desperately with the forces protecting the supply lines, and they may yet prevail. If they do, and if the First Cavalry stops the attacks from reaching Afula and Jerusalem, then I think the enemy will begin a general retreat. If they don't, they will be annihilated.

"We've lost contact with most of the units in the front lines along the Jordanian border, but we know from reconnaissance flights that they are fighting a desperate battle with the enemy's rear. We can only surmise what is happening, but I think that small unit commanders have taken charge and continue to fight even though their command infrastructure has collapsed. You can be very proud of those boys, Madame Prime Minister, very proud."

The Prime Minister stared at Gen. Weiss and said nothing, her eyes tearing as she walked out of the command center. She made it to her apartment before collapsing. "I *am* so proud of them, so proud of all of my country. Please God, deliver us from this nightmare." She murmured her prayer softly, and then indulged a moment of despair before she collecting herself to reenter the command center.

DELIVERANCE

Ariella awoke with a start. *Oh my God how long has it been?* she wondered. Forty-five minutes, by her watch. She must have fallen asleep. She was very angry with herself and cursed her weakness. She marched down the hall to the command center to get an update. As she entered, she could hear cheering from the officers and staff in the room. She glanced up at the screen and saw many opaque red symbols blinking, and other red symbols moving back towards the Jordanian border. She also saw blue symbols moving east in pursuit.

"We did it Ariella! We did it! They're in a disorganized retreat along both salients, dropping their equipment in their panic." Gen. Weiss shook Ariella's hand violently and then gave her a bear hug. He stepped back, beaming a tired but jubilant smile, unembarrassed by his outburst of emotion.

Ariella looked serene. She took both of the general's hands into hers, leaned forward, and kissed the general on both cheeks, slowly and with great feeling. She allowed herself a soft smile and mouthed the word, "Shalom."

Her people were saved.

LIES A WOLF AMONG THE SHEEP

The members of the Islamic Defense Council (IDC) were severely depressed to hear that their forces had been beaten back after so much success early in the battle. They began arguing among themselves and blaming each other for the various tactical setbacks, any one of which might have turned the tide of battle in their favor. They cursed the Americans and their new technology and debated whether they should reinforce their troops and mount another attack against Israel.

With a calm, authoritative voice, a young man, barely thirty, a respected sheik and seer, addressed the leaders: "We've taken back lands that were once Arab lands. In time, we'll re-conquer the rest and drive the Zionists into the sea, along with their apostate American allies. For now, we'll hold our positions and concentrate our efforts against the godless American general approaching our shores. When we've secured the holy lands against the infidels, we'll finish with the Zionists."

MR. PRESIDENT

When the Senate convened on Wednesday morning, the newscasts were abuzz with the story of the desperate battle for the survival of Israel. This news, while expected, was a call to arms for the men in the chamber. Twenty-four senators and forty-seven governors sat down with a heightened sense of immediacy. As the Senate was brought to order, the provisional majority leader asked the assembled men and women to submit names for consideration for the next President of the United States. By the rules they had established, each name would have to be seconded by at least three others to be considered. By the same rules, if there were more than two candidates, the winning candidate would have to win at least fifty-one percent of the votes, or a second vote would be taken between the top two vote getters.

Sen. Fitzpatrick's name was entered into consideration by one of the "surviving" seven and quickly seconded by four other senators. Governor Jones' name was similarly submitted and seconded. Governor Tilden, controlling both votes from his state, submitted, and seconded, a third candidate. Another two-vote governor provided the additional seconds needed for this third nomination.

Tilden's nominee, acting Secretary of State Stanley Donner, was a surprise to Sen. Fitzpatrick, who cheerfully surmised a rift between Tilden and Jones that would redound to his favor.

As the roll call began and Governor Jones cast the first two votes for Secretary Donner, Fitzpatrick knew instantly that the fix was in. As one smiled and the other scowled, both Jones and Fitzpatrick knew that Donner would win. Fitzpatrick held his rage in check but threw a savage glance at Greg Bokert, who remained expressionless, unafraid of his boss's rage.

The final vote was sixty-three for Secretary Donner and thirty-seven for Fitzpatrick. Secretary Donner was sworn in as the next President of the United States.

Greg Bokert met Sen. Fitzpatrick in his chambers after the session.

"What do you have to say for yourself, Greg?"

"You're an idiot, Senator—unqualified to lead this nation, or to lead a Boy Scout troop, for that matter. I set this up, set you up, and I'm proud that I did. I've joined the Templars, and I take leave of you for a far better job as a private in the Third Templar Division."

THE SCOFIELD PROBLEM

"Congratulations, Mr. President," Gen. Scofield greeted President Donner via a secure video link over satellite. "I was pleased to hear of your election and look forward to meeting with you as soon as I return."

"Congress has demanded that I order your forces to disengage from this conflict and return to the United States immediately. They also demand that I replace you with Adm. Franklin and have you sent back to the United States in chains to face a general court-martial, followed by disembowelment or hanging."

"Which would you prefer, Mr. President?"

"Well, I know you are a glutton for punishment, General. I'm leaning toward disembowelment, but would probably toss in a quartering just to make sure you receive the full measure, so to speak."

Both men laughed. Donner explained that the current Senate, consisting of only twenty-four senators, would remain *Lake Fitzpatrick* until the rest of the governors' senatorial nominees were sworn in. Fitzpatrick wasn't likely to move quickly on their induction, and Donner's life would be a miserable one in the meantime. But the House was friendly enough: they'd promised at least twenty-four hours of honeymoon before turning on him.

Scofield described the tactical situation to the new president, who was keen to hear about the Israeli victory in the first round of the battle. His brief was as current as it could get, unlike the newscasts still predicting doom for Israel. *Thank God,* the new president thought when he heard Scofield's forecast that Israel would survive and that the stalled attack would remain stalled as the general grabbed the neck of Islam and started shaking.

"How many casualties, General?"

"The Israelis are not certain yet, but it's bad—very bad. Survivors will show up in the coming days, but preliminary counts have the dead at eighty thousand and climbing, with another twenty thousand civilians dead. The wounded number in the tens of thousands and may grow to over three hundred thousand, including at least one hundred thousand civilians. They're currently reporting more than one hundred thousand missing, but many of these will be stragglers separated from their units.

"Most of the carnage occurred in a matter of hours, Mr. President, with only about fifteen percent of the casualties in the days preceding the battle. The manpower of the Israeli forces has been cut in half, and their losses are disproportionately higher in their first-line units. They lost eighteen hundred tanks, one hundred and twenty aircraft, and every naval fighting ship they had. Their overall casualties represent about one in ten Israelis."

"What are the estimates of Arab casualties?"

"Roughly two times the dead and an equal number of wounded. They may have lost as many as three hundred aircraft and twenty-six hundred tanks. It's difficult to determine. We do know that they lost every naval vessel they had in the Mediterranean Sea. You can thank us for that."

"What are *our* losses, Mike?" President Donner dreaded the answer to this last question. He would have to explain to the American people why American men and women were dying in Israel, defending Israeli ground.

"The First Armored Division suffered one thousand, three hundred and forty-seven dead; two thousand, one hundred and ninety-two wounded; and one hundred and fifty-seven missing. That represents a twenty-percent casualty rate, Mr. President. Their commanding general was killed in action."

"And the First Cavalry?"

"They had a rough go of it. They suffered two thousand, four hundred dead; three thousand, six hundred and twenty-seven wounded; and four hundred and fifty-nine missing in action. Their lead forces were nano-armored, but the rear was committed to battle without armoring. The Arab air attacks were very effective. This represents nearly a forty-percent casualty rate."

"Was it worth it, General?"

"If the army had not sacrificed so many brave soldiers, the Arabs would certainly have overrun the Israelis, who would most assuredly have launched a full-scale nuclear attack against their enemies as their last act of defiance. It's probable that we would not have been able to contain the conflagration, and we would have ended up with global thermo-nuclear war."

"That's not what I'm asking."

The general understood what the new president was asking and respected him for asking such a probing question. If the general had not decided to invade and destroy Mecca, and had not announced this to the world, then 100,000 Israelis, 200,000 Arabs, and 3,800 American soldiers might still be alive. Worse than the dead littering Israel was the possibility that his decision could have caused a global nuclear war. Could any objective justify such risk?

"I made that decision alone, so these dead are blood on my hands. Three million dead Americans already litter our land, with many more yet to die on both sides. That's a tough equation to balance, Mr. President. I can only look to history."

"Go on, General."

"Great conflicts between civilizations require an organizing martial philosophy to motivate the masses to sacrifice life and limb. To defeat an enemy who has such strong magic, you must tear down the martial philosophy so that the survivors reject it and accept a new one. That requires that you completely tear down the civilization that spawned it. The Nazi form of fascism died only when the country that spawned it was utterly destroyed. It was the same with Japan. Only with the burning to ashes of Japan was the fallacy of the divine Emperor and the Bushido code finally rejected, along with their suicidal adherence to their martial philosophy. It's like a psychotic episode that rewires a person's brain, except that it happens to an entire civilization collectively."

"But *everyone* could die if this goes nuclear. By your own admission, we came close to that today. The equations require that *risk* be balanced, too."

"The core martial philosophy that has infected Islamic thought is based on Wahhabism, which is named after the Saudi cleric Muhammad ibn Abd-al-Wahhab. This philosophy allows Islamic warriors—who truly believe they are sanctioned by God—to do terrible things like murder three million Americans with nuclear weapons. As a powerful unifying philosophy, it also has great patience. Read the writings of Osama Bin Laden, and you'll see how patient they are willing to be. Their philosophy requires them to rule the world, in time... to convert the world to the Islamic faith and to the rule of Sharia law. The hardliners understood this when they called for a nuclear retaliation against the whole of Islam. The so-called 'death from a thousand cuts' is a philosophy that's meant to whittle a stronger foe down until he is weaker than you, and you are ready to strike. But the process is accelerated when the cuts are nuclear attacks."

The new president completed the thought for Scofield. "We have to destroy Islam itself because it's so poisoned by the cancer of Wahhabism that it cannot be saved. Burn it to ashes, or we'll be destroyed."

"In my view, yes, the hardliners were willing to jump immediately to nuclear conflict, to use our considerable advantage while we still had it. They unwisely thought that they could contain such aggression. My plan is certainly safer than unbridled nuclear attack, but both plans presuppose that we'll be destroyed and subjugated if we don't prevail. Without that core belief, you could never balance the equation in favor of the destruction we are wreaking. Certainly the risk would be too great, even if the carnage were acceptable. If you don't believe that this war is unavoidable, then you should reconsider my disembowelment."

"I'd already come to the same conclusion, General.

After a pause Donner continued. "You know I have a background as a historian."

Scofield nodded.

Donner continued, "I've been reading the Qur'an to try to gain insight into the teachings of the Prophet and to understand Wahhabism. I was struck by the contrast of the writings of the Prophet and Jesus Christ. I now understand why Wahhabism has become so deeply rooted in Islam. A detailed study of the book leads directly to Wahhabism. It's the correct interpretation, in my view. Frankly, I was shocked that I came to that conclusion given what I've read in the press and in secular research papers. Going to the original text was illuminating.

"My conclusion was that the Qur'an is, at its core, a martial code, a formula for world domination sanctioned—no, demanded by—Allah. The New Testament and Christ's teachings are rooted in a philosophy of love, which is antithetical to Islam," Donner concluded. "This conflict was predestined."

"I agree, Christ preached love as the core doctrine of worship and faith. Muhammad instructed his people to make war to expand the faith," Scofield concurred. "A sharper, clearer contrast cannot be found."

"We are of the same mind then—victory at any cost. Where do we go from here?" Donner said.

"I'll finish my job, and you should endeavor to keep the vultures off my back. Strength and honor."

Stanley Donner had stood with him in Afghanistan against the Iranians. A master sniper, he had saved Scofield's life many times while providing overwatch for the Lions.

"And a good death, my old friend. Just make sure you aren't one of them."

TO BEG FORGIVENESS

Told to stay with the Prime Minister and protect her with his life, against all enemies, Capt. Buckhold had lied to Ariella, telling her he had been ordered to the Israeli battlefield. But he had stayed by her side through it all, only sleeping when she slept, which wasn't often. His heart bled with hers as she collapsed in tears in her private apartment. She was utterly alone, beset by ten nations sworn to her people's destruction. The reports of the casualties rolled in like an endless nightmare. The numbers were staggering.

Capt. Buckhold had come to adore Ariella, even though she was twenty years his senior. He wept for her and worried that she would not be strong enough to see this through. Putting a call in to Gen. Scofield, he was surprised to find himself connected almost immediately. The captain started to explain what was happening to Ariella, what she was going through, but could not finish. The general waited patiently for Buckhold to regain his composure. When he had, Buckhold asked his general to come to Israel to lend Ariella his strength. Buckhold was one of many Templars who felt that his general was anointed and could heal with his touch.

"Michael, thank you for coming here to see me," Ariella greeted Scofield. "I can't tell you how much it means to me." Sitting in her underground bunker's private apartment, sipping coffee with the general, she looked exhausted, a thousand years older than days before. Ariella was the leader of her people at the darkest point in their modern history, a history begun in the late nineteenth century with the planting of a small Zionist seed in the infertile ground of Palestine. Zionism ended the Diaspora, later given impetus by the horrors of the Holocaust.

"I've come to ask you to wash the blood from my hands with your forgiveness," Scofield began.

"I don't know if I have the strength for that, Michael. Your nation has asked so much of mine these last hours. We survive for now, but I fear we don't survive much longer. My armies have bled more than I thought they could. The dead are stacked in piles, waiting to be buried in mass graves, their number not easily counted."

"We won't allow the Jewish nation to be cast into a second Diaspora. When I've drawn the enemy to me, when I've destroyed him in his home, I'll return to destroy the armies that lie in wait for you. We'll soon be entering the Red Sea, and our guns and missiles will be joined with yours. Already our Air Force bombers are wreaking havoc on your enemy's ability to regroup and threaten your lines. Without air superiority, they cannot attack, and your air force, combined with our naval air resources, won't allow that. You must take strength in what you have and persevere."

"To forgive you, Michael, is to forgive God for his abandonment of my people. I can do neither."

"I understand despair, Ariella. It falls to you to suffer on your soul the stain of this bloodletting; you are its leader, as I am its architect. Forgive me. Forgive yourself."

Ariella slumped in her chair and began to cry. Scofield moved quickly to support her as she fell from her chair into a fetal position in his arms. Her cries turned to sobs. Scofield embraced her, giving her time to pass through this moment of crisis. Her tears dried, and she lay motionless in his strong arms. He touched his hand to her forehead and willed his strength to her, asking God to soothe Ariella's grief. Ariella felt warmth on her forehead that was unnatural but hypnotically soothing. She fell into a deep and restful sleep.

RUNNING THE ENTRANCE TO THE RED SEA

The crisis in Israel demanded a shift in Adm. Franklin's timetable. He sent the majority of his sea jets into the Red Sea to make war on any surface vessel stationed there. Frigates, destroyers, and submarines were dying en masse in the small sea, with the Arab air forces helpless to stop the carnage. They could not kill what they could not see.

To test the strategy of the air and naval forces defending the Red Sea, Franklin sent two nanoclad frigates into its mouth. With the entrance barely ten miles wide, the frigates were immediately beset with volleys of surface-to-ship missiles. The nano-armor protected the ships for a few missile rounds, but could not withstand the cascading strikes and eventually failed. Both ships erupted in violent explosions.

The nano-armor, once penetrated, had the undesirable effect of containing the explosion inside the ship's infrastructure. This greatly amplified the explosive effect, together with the cascading secondary explosions within the hull. The ships blew to pieces, killing all aboard.

Saying a short prayer for the dead, Franklin watched on his screens as his nano-armored aircraft closed on the shore batteries that had launched this hell upon his navy. When he was sure the batteries were dead, he sent another frigate through the entrance. It, too, was attacked with volleys of surface-to-ship missiles, from locations further inland. Franklin sent in his aircraft to destroy the killers of his third ship. He then ordered sea jets to scour the Red Sea entrance for sensors alerting the enemy to a ship's presence. He also sent a fleet of B-52s to carpet-bomb the next likely source of missile batteries.

Franklin was considering his next move when the sea jets reported discovery of a set of listening devices placed at the bottom of the Red Sea's entrance.

The devices had floated tiny wires up to detect any ship passing above them. The sea jets engaged and destroyed the listening system.

Franklin then sent a destroyer from his forward carrier task force through the sea entrance. It passed though unchallenged. Stationing this ship just inside the mouth, he ordered the sea jets to seek out and destroy all listening devices inside the Red Sea, a time-consuming task—Franklin wasn't happy to be interrupted by Scofield in the midst of it.

"Who do you think helped the Arabs plant the listening devices?"

"You tell me. You have all the whiz kids. I'll see if we can recover one or two and do some analysis. We'll have to get back to you on that."

"I have spies everywhere, but I didn't get wind of this. The only blind spot I have is Russia; my invisible spies have been detected and silenced there. The Russians seem to have found a way to see through our nanocloaking technology. That could be a big threat."

"The sooner we know, the better. They may have just laid a trap for one of our carriers, letting the last ship through as a ploy. The listening devices could be a decoy. If they can see a Templar Knight spy, then they can see a cloaked ship in the water."

"I hope you're wrong, Admiral. I'll get Mouse on it right away."

"Can she also program the nano-armor to sense when an attack has penetrated a ship's hull, and allow an explosion to dissipate rather than cause a catastrophic explosion? My boys can keep a wounded ship afloat and keep fighting as long as it doesn't implode."

"She was horrified when she saw the video of the attacks and went to work on it immediately. She is already sending updated algorithms out to the ships' armor. It may take some tweaking, but she'll get it right."

"Tell her she has my undying gratitude."

MARK'S JOURNEY

Maria had not seen Sally Furlong for over two years. Having just arrived on the flagship, Sally was tasked by Gen. Tommy Marks, Scofield's divisional executive officer, with setting up and running the divisional headquarters' battalion aid station. Sally Furlong and Mary DuPont were both ex-Navy corpsmen who had attended medical school after serving with Scofield in Afghanistan and Iran. The rest of the medical staff was still Navy—the tradition of Navy corpsman and doctors was deeply etched into Marine Corps tradition. The two Templar doctors had been given special dispensation to join the Templars as Marine officers because of their special knowledge in Scofield's genetic elixirs.

Sally had a long history with Maria, having patched her up back in the days before they wore nano battle suits—when bullets tore flesh and shrapnel maimed. When they'd first met, Sally had been a corpsman in Afghanistan, and Maria, grievously wounded, was just another Marine sergeant dying on her watch. Sally had judged Maria's wounds to be fatal and left her to attend to other wounded, as called for in triage procedures. When she had come back, expecting Maria to be dead, she'd seen a gruff Marine colonel kneeling over her, praying, touching her forehead and wound, oblivious to the blood seeping down his own back. Sally had seen her first miracle that day when Maria bolted up from unconsciousness. The image of Maria with her bloodstained lips and teeth was burned into her consciousness for the rest of her days. Sally's long journey with Maria and Scofield began that day.

Sally loved Maria. It was impossible not to fall in love with her, with or without carnal desire—she had that effect on people. To Sally, whose love for Maria was as profound as it was suppressed, this was a mystical bond.

"I haven't seen the General since I returned," said Sally. "Is he here?"

"He is visiting a new friend in Israel, but should be back within the hour. He will be delighted to see you. How is your newest project going?"

"It's going fine. Mark survived the experience and is doing well. We had to condense the whole process into eight hours to allow him time to heal before we land at Jeddah. His faith led him through his ordeal."

THE RED SEA

The passage of the first carrier battle group through the opening to the Red Sea was uneventful. They had waited until nightfall to give the sea jets time to check for more sensors, trip wires, and mines. Finding them around the port of Jeddah, Franklin surmised that the sensors were probably not a ruse and that his ships were still invisible to his enemy. Nonetheless, he took all necessary precautions to protect the carrier, including flying helicopters next to the ship to intercept any missiles. The helicopter pilots, running a skeleton crew of two, prayed that the nanocloaking was working.

Once they arrived in the Red Sea, they started air operations immediately, expecting all hell to break lose.

It did. As Franklin had predicted, the Islamic bloc deployed only a fraction of their aircraft in this initial phase. Still, at over two hundred planes, it was a substantial force. The Arab air commander tried new tactics to crack America's technical advantages, including ramming ships with low-level suicide jets and launching blind missile attacks against likely target areas. Within three hours, however, the United States Navy dominated the southern Red Sea air operational area.

The Americans felt admiration for the many Muslim pilots sent to their deaths that day, courageously undertaking their missions while being mercilessly attacked by nearly invisible American aircraft.

Franklin's forces had sustained three hits against two frigates, which were damaged but still underway. One destroyer was lost when a suicide jet struck it—and was then attacked mercilessly as it trailed oil in the water. The carrier *Enterprise* was hit by two missiles—two good guesses—but sustained only minor damage and remained in the fight.

The carrier was so busy launching and recovering planes that Franklin knew he had to either establish a land-based air facility or bring a new carrier into the Red Sea. He ordered a contingent of Templar Knights to be landed at a small island off the coast of Sudan, near the city of Muhammad Qol, to secure an airstrip. A small, amphibious ship offloaded a group of Seabees, who went to work building a rapid deployment airfield and air-operations support facilities. They had it operational twelve hours later. This island base put aircraft directly opposite Jeddah across the Red Sea, only 120 miles striking distance from Mecca. The airbase would be used for bomb refit and refueling operations, with other maintenance left to the carriers. The extra space also allowed

Franklin to send in more carrier planes than the *Enterprise* could normally handle by itself, and leave his two remaining carriers out in the open seas of the Indian Ocean until he was sure he had full control of the Red Sea.

FAITH OR DESPAIR

"We'll be landing soon," Scofield announced. "This will be our last quiet meal together, ladies. I suggest that we feast first, then sit back and talk philosophy."

Maria, Mouse, and Scofield dug into the meal like beggars at a king's banquet. Scofield finished first and performed his usual post-meal ritual, releasing a weapon of mass destruction. Scofield laughed as his two guests cringed. They smacked him on the arm, then started laughing. Caught with a mouth full of dessert, Maria spewed her food across the table, sending everyone into hysterics. They laughed loud and long, falling to the floor clutching their aching sides.

When the insanity subsided, they remained on the floor, smiling at each other, quietly meditative. Maria held her hand out to Scofield, and Scofield extended his hand to Mary Louise. It was enough just to sit there and be in the moment.

After a while, Scofield helped the two women up off the floor and led them into his den, where he fixed each a drink.

"What shall we talk about, ladies?"

"How about your theory for God's indifference to our suffering?" Maria suggested, hoping to attack that myth.

"We could discuss the Two Great Mysteries. I'm the birthday girl, so I should get to choose," Mary Louise asserted. "Let's talk about the Second Great Mystery."

"What are the Two Great Mysteries?" Maria asked.

"It's a debate that Mike and I have been having for the past two years, while you've been hibernating with your sisters. The First Mystery concerns the necessary conditions for the creation of the universe. The theoretical physicists have been so stumped by this question that they've turned to mysticism to find acceptable answers. The latest craze is string theory, multiple dimensions, and parallel universes. With these constructs, virtually any mathematical system can be shoehorned in—presto, problem solved. Mike and I both believe that this is too convenient, so we've been making our own postulations. Of course, we've been unsuccessful. Since neither of us has the background for it, we content ourselves with making fun of the physicists and choose to ignore Mystery Number One."

"Why not God?"

"Well, both Mike and I, as engineers and amateur scientists, like to find solutions first in science. We both believe that God did his work through nature, not against it—that God *is* nature. We believe that in time, science and religion will be reconciled, and that God will be revealed through the mystic calculus. In other words, everything is a calculation, and the professor is God. Meanwhile, Mystery Number One is too hard, at least for us, so we skip to Mystery Number Two."

"I'll bite. What is the Second Greatest Mystery?"

"The origin of life," Scofield added.

"The best the biologists can come up with for the origin of life is that a bunch of random amino acids decided to get together and party in just the right way to form a simple cell, and this led to more complex cells, and that eventually led to us. Of course, many biologists don't accept us as an evolved life form, preferring instead to think of us as an earthly *virus*," Mouse said dismissively.

"We have an advantage, Munchkin. We've seen the program. The random formation of life in the primordial soup is simply a theory, nothing more. Don't take it personally," Scofield scolded. The general liked to debate science with an open mind, allowing all possibilities, but the brilliant Mouse could be stubbornly opinionated.

"But it's absurd on its face, Scofield. The contention that two hundred or more amino acids would self-organize into something as complex as a cell is preposterous without evidence to back it up, and there is none. Even if such a miracle did happen, the resulting cell would have to learn to replicate itself before it was killed in the toxic primordial soup. The likelihood is that the base cell would have to be 'self-invented' many billions of times before it learned to make babies."

"Okay, so how did God do it?" Maria asked. "Since we know He did."

"If we just wave a wand and say God did it, we are falling into the same mystical trap as the physicists, though I'm sure they would take offense at my characterization," Scofield said. "I want a better explanation—one that incorporates nature."

"And you are stumped?"

Mouse and the general nodded.

"You see evidence of the miracle of God's creation all around you, you have personally experienced divine revelation and the miracle of healing. You have healed me for God's sake. Yet you cannot accept that God breathed life into this world through miraculous intervention? Who is being close minded?" Maria looked toward Scofield.

"The improbability of life generating spontaneously from stardust is so great that, for now, it defies explanation," Scofield said, ignoring her point to continue his narrative. "It's so unbelievable that if I didn't see all the varieties of life on this planet, I would not believe it possible. Man is the *existence proof* that a theory exists.

"When you inject the knowledge that we've discovered—that the genetic code is, *in fact*, an organized computation, with complex algorithms embedded into that calculation—then the odds radically skew against random chance."

"Another dead end, then," Maria asserted. "Leaves you with the choice of accepting despair or choosing faith. So why not choose faith?"

"Because it's too easy," Mary Louise said. "One of us blindly and stubbornly adheres to the idea that the existence of God *must* be proven through an understanding of nature. For my part, I've looked into the abyss and chosen faith, because there is no other explanation available. Ockham's Razor."

"*Lex parsimoniae*," Scofield grumbled.

"We choose faith, and you willingly choose despair, against all logic," Maria summarized, glaring at Scofield. At last she understood his core sickness, the self-destructive nature he had finally revealed to her. Scofield looked uncomfortable under her gaze. It made perfect sense: he had been seeking his own destruction from the beginning. Somehow she had always known, but she didn't want to believe it.

Mary Louise understood why someone might seek his own destruction, having suffered for so many years with her own demons. She saw what Maria saw, but, unlike her, had no trouble making sense of Scofield's contradictions.

"You have all these advantages, my love. Three women that love you: Sarah, Mary Louise, and me—not to mention four daughters and six grand-daughters. You have felt the presence of God and have been visited by angels, an experience denied to most of the faithful. You have seen God's reflection in nature, unlike any before you...yet you still will yourself a nonbeliever. Why do you choose despair, Michael?"

"The Third Great Mystery," he said.

"Is that why you distract us with this charade about discussing philosophy while our troops prepare for war and death? Just another fucking day, eh Scofield? But God forbid we should talk about the real elephant in this room." Maria paused. "Are you glad that your death draws near?" Maria's voice had begun harshly, but it trailed to a whisper as she spoke these last words.

"You know the answer to that, Maria," Scofield said.

"No, Mike, I don't. Enlighten me."

"Death holds no mystery for me. To return to the womb of beauty, comfort, and peace waiting in heaven, even if only for a moment, draws me toward death like the moth to the light. My duty, my family, and the few friends I have left sustain me here in this world. The rest has lost its luster. This long war has beaten me down."

Mary Louise was crying. She leaned over to her mentor and hugged him with all her strength. "I'll be lost without you."

"As will we all," added Maria.

"Maria, you must assume my duty after I am gone. You must carry on."

"When you die, I die. I am not cut from your cloth. I don't have the de-tached cruelty needed to execute war," Maria said.

"No one is born with that. It's a curse...an acquired skill born from years of battle." The general paused. "You must do what God asks of you." Scofield was looking at Maria as a commander to his subordinate, expecting her to obey.

"I am not worthy nor prepared. I would fail." Maria said. *Please don't make me swear this oath, my love. Please.*

"You have always been the one, Maria. Swear to me that you'll do what's necessary after I'm gone."

She looked to him, hoping he would relent but his stare was answer enough. "I swear, General. On my oath as a Templar and a Christian warrior, I swear."

Scofield brought the two women into a circle, each holding the others' hands. "Mary Louise, you must swear a similar oath to Maria. I need to hear the words."

"I swear an oath of absolute allegiance to Maria should you fall my liege," Mary Louise said. "To the death I swear, so help me God."

"I love you, my dark angel. No matter where your demons take you, I'll have faith enough for both of us," Maria said for both of the women.

OPERATION RATTRAP

Brig. Gen. Tommy Marks and his two Templar escorts had shadowed the mujahedeen fighter for ten hours. He was a cautious fellow, constantly doubling back and laying ambushes for anyone who might be following him. He couldn't afford to be followed to his secret destination. He was the only outside contact for Sheik Abdullah Caliph al-Rashid, the object of a manhunt even more intense than the one that had scoured Pakistan in search of Osama bin Laden. When bin Laden was their prey, the infidel's technology had been good, but not omnipotent. Now the mujahedeen had to contend with "ghosts." Though he scarcely believed the stories, so far beyond his experience, he had been told of the infidel general's ghost warriors.

After walking for ten miles to travel seven, the mujahedeen fighter finally ended his meandering path at the mouth of a cave. There were four heavily armed guards at the entrance, more within, and more without.

The fighters were having dinner and relaxing when Marks stepped into the cave and uncloaked. The shock on their faces was palpable. Two men stationed outside the entrance, suicide bombers tasked with killing the sheik to prevent capture, were decapitated by cloaked Templar swordsmen before they could respond. A flash of blue light, and their heads tumbled to the ground. A guard trying to reach the bodies and detonate the bombs was also separated from his head.

Gen. Marks introduced himself to the startled group as Gen. Scofield's assistant divisional commander and asked for a temporary truce so that he could discuss the general's terms for battle. The sheik readily agreed, happy that he wasn't being dragged to an infidel prison.

"Gen. Michael Scofield, United States Marine Corps, challenges you to a personal duel to the death," Gen. Marks said in perfect Arabic. "He requires that you meet him at a site of his choosing near the sacred Kaaba. You may choose a champion to fight for you. The battle will be with swords to the death. If Gen. Scofield kills your champion, you, Sheik Abdullah Caliph al-Rashid, will also be killed by the general. If you kill the general, then all United States forces will withdraw from the Middle East, and the sacred sites and the cities shall be left unharmed."

"If I don't agree?"

"The general will soon be standing at the foot of the Kaaba. He will defile the shrine and lay waste to the two holy cities if you don't agree to his terms. He will televise this meeting to the world and call you a coward for leaving Islam to the infidels."

"If I am martyred by this devil, will he not still lay waste to the Kaaba and the holy cities? I expect this usurper's assassins to lie in wait and murder me, coward that he is."

"The general has no interest in the holy cities or your sacred black rock. He wishes one final battle to decide the issues between us. You'll be required to bring with you every one of the one thousand fighters named on this list." Marks handed the sheik a computer disk, "Each man on this list must stand in the first two rows of your battle line holding a token genetically coded to identify him. Should a single man not appear, the offer is void, and your cities and shrines will be destroyed. Your men will be given safe passage to the battlefield once the general has made camp outside Mecca. Until then, we'll kill any and all who move towards Mecca. While you are in transit, a truce shall be extended, and we'll wait for the final battle without harm to you or the Islamic armies during that time."

Mulling the proposal, the sheik knew he was trapped. He could not permit the general to dictate the method and terms of his death, but it would be disastrous to allow defilement of the holy sites without showing courage against the enemy. To agree was certain doom; to refuse, humiliation. The sheik resolved to die bravely as a martyr for his people.

"I'll meet the general on his terms, and I'll bring a holy warrior as my champion to fight this devil. The muslim world will see that your weapons make this 'duel' nothing more than murder. They will know the truth of it. If we die, we do so with God's will, and our deaths will inspire the faithful to defeat the Great Satan."

We have a weapon, sent to us from God and he will avenge my death. The Caliph pondered how the sheik would kill this apostate.

Unknown to Sheik Abdullah Caliph al-Rashid, these words—indeed, the entire confrontation just concluded—had been broadcast live around the world by video feeds from the Templar Knights.

CHAPTER SIX
THE FINAL CONFLAGRATION

JEDDAH

"In the last two days, we've solidified our domination of the Red Sea and the air around it," said Adm. Franklin, beginning the briefing. "Our interdiction of the forces closing on the Mecca operational area has proceeded without major challenge from the Islamic air forces. They appear to be holding the rest of their air force in reserve for the final battle. In response, we've shifted our Air Force bombers, targeting the outer airfields from which their attack will be launched. Thanks to our pounding, they're running out of airfields in the immediate vicinity of the Red Sea. Navy aircraft are being used in the interior area of operations. We have two aircraft carriers in the Red Sea, and three land-based air bases on three isolated Red Sea islands to support air operations. We have naval aircraft from the Navy and Marine Corps team, as well as every tank killer aircraft in the Air Force and Army inventory that we could scrounge. We're ready to land the landing force."

"That's excellent news, Admiral!" Scofield affirmed. "Paul, would you brief us on the amphibious operation."

"All of the ships are in theater," Adm. Wainwright began. "The landing plan is complete and has been distributed. We've warned the residents of Jeddah, Mecca, and the surrounding communities to evacuate immediately. We specifically told the Jeddah residents that we intend to reduce the city to rubble. Unfortunately, they haven't complied. The Islamic Defense Council has ordered all Muslims to stay and fight—even the infants, presumably. Jeddah is a city of over three million people, and Mecca has approximately one-point-seven million, and they've been bolstered by two million men from the armed forces of over twelve nations. The bottom line is that civilians are being used as shields, and they will die by the tens of thousands. To help civilians flee the city, we intend to interdict three main Jeddah exit points that have been guarded by Saudi military units to keep civilians penned up in the city. That will at least give some civilians a chance to flee. We'll be able to neutralize the army units, but civilians may still die at the rate of two-to-one during the initial assault through Jeddah.

"The Templar Division will land serially, docking armored and cloaked amphibious ships at the main dock. The units will disembark while cloaked and move along Falah Street to the main assembly area, a large open space near the Prince Abdullah al-Faisal Stadium. The Templar units will remain cloaked until the arrival of the general's staff, with Gen. Scofield assuming operational control of the division at that time.

"During the landing and assembly period, we'll avoid direct confrontation by skirting the main parts of Jeddah. Aerial bombing will provide cover. The city will be preoccupied with its destruction. By the time the division steps off on its march to Mecca, Jeddah will be a ruin.

"While we're landing the Templars at Jeddah, we'll also be landing the army units near Ash Shaiba, an oil-loading and storage facility about seventy miles south of Jeddah. These units consist of the nano-armored tanks, personnel carriers, artillery, and air-support units the Army has lent our enterprise, plus various supply units. They will offload at Ash Shaiba and move to an area inland along Highway Five to a defensible position south of Mecca. This path will be our main resupply route, and the staging area for our armor and attack-helicopter support bases. We've been supplied with mobile nano-armored shields to protect our base, and one thousand Templar Knights will be vectored to the base to protect it using the First Battalion, Third Regiment. First Battalion will peel off from the main Templar column that will be moving to Mecca via the Alharamain/Jeddah-Mekkah Expressway. First Battalion will meet the army units as they approach the assigned base area."

"Melissa, take us through the Templar Division plan of attack," Scofield said to his chief of staff.

As Col. Melissa Franks, USMC, rose from her chair, the entire assembly clapped and whistled. Every man and woman there understood how important the Pakistani operation had been to the success of the operation. Col. Franks nodded her thanks.

"The Templar Division will assemble in the area just west of the Prince Abdullah al-Faisal Stadium," she began, "in a column of regiments, with each regiment in a column of battalions. Each battalion will be arrayed in a column of companies, with each company arranged in a fighting square ten Knights wide by twenty-two rows deep. The formation will extend over four thousand feet in length from tip to end. On command, we'll de-cloak the division and begin our march down Falah Street to the Alharamain Expressway. We'll march and run the thirty miles to Mecca, fighting and defeating any enemy forces we encounter along the way.

"Our forces should provide sufficient bait for the tank-killer and attack aircraft that will accompany us on our jaunt. The aircraft flying and hovering over our formation will accomplish the majority of the killing. We'll have four hundred attack helicopters, two hundred A-10 armored warthogs, and various Marine Corps attack aircraft escorting the division. The aircraft will cycle through the battle, replenishing weapons and fuel in coordinated waves so that the division will have approximately twenty-five percent of its air assets fueled, on station and ready to fight, at all times. The air assets will remain cloaked. The Navy should intercept any enemy attack aircraft before they reach us. The Navy has established an outer protective ring around the approach line of march that will extend for three hundred miles in every direction.

"Remember, nano-armored warriors are protected against many threats, but not all. Any armor-piercing round equal to or larger than a twenty-millimeter shell that has thermal penetration characteristics *will* penetrate the armor and kill the Marine wearing it. It requires a direct hit, but it will kill. In time, the enemy will figure out that we are not invincible and will adjust his tactics to kill us. All air assets should focus their efforts on tanks, anti-air twenty-millimeter cannons, artillery, and any other gun capable of killing a Templar Knight. We'll take care of the ground forces.

"We'll stay on the Alharamain/Jeddah-Mekkah Expressway as we enter the so-called 'third ring' in our approach towards the al-Masjid al-Haram Mosque, the 'Holy Mosque.' The freeway turns into the Umm Al Qura Road, which leads directly to our objective.

"Along the march, and especially when we reach Mecca, the order is, kill anything that moves, without regard to collateral damage. We have a number of sacred sites that will be marked off-limits—the Holy Mosque and the Kaaba being two of them. However, if it isn't marked, kill it. Understood? We launch the final phase attack against Jeddah tomorrow at dusk and land the Templar Division during the early morning hours. We'll cross the line of departure from our assembly area at daybreak the next day."

Gen. Scofield stood to give his summary: "Ladies and gentlemen, you are all battle-hardened veterans of this long war," he observed. All of you have lost comrades in battles against the Islamic radicals, and most of you have lost loved ones in the attacks back home. We've come to seek a reckoning for those dead, to answer their suffering by inflicting a grievous wound to our enemy. We've risked much to get here; our ally Israel has risked and suffered much more. We've drawn out our enemies, and they will present themselves to us soon at a battlefield of our choosing outside their holy city of Mecca. Prepare yourselves and your charges for a series of one-sided battles far bloodier than any before.

"We must be prepared to wrest all hope from our enemy's souls. In order to crush this evil philosophy, we must bring the Islamic civilization to ruin and demand their commitment to peace. If we fail to achieve that goal, then the next time we meet them it will be with the hellfire of nuclear war. Your strength and willingness to slaughter the enemy, even the innocents the cowards hold in front of them, are the price you must pay for a long-term peace and the long-term survival of our nation.

"I am proud to lead such men and women in this great struggle. Strength and honor!"

"And a good death!" Even the Navy commanders echoed the Templar oath.

SARAH, MY LOVE

"General, it's Sarah." Capt. Weaver leaned over to tell Gen. Scofield that his wife was on the satellite phone. Scofield and Weaver were in the command center following the progress of the air campaign.

"I'll take it in my quarters."

"Roger that."

"Hello, Sarah, how are you? What time is it there?"

"It's three in the morning; I've been watching the news, following the progress of the war, and worrying about you, my love." Sarah Scofield was at their ranch in the Shenandoah Mountains with their three daughters, their husbands, and their six granddaughters. They were under armed guard by state police on the perimeter and a squad of Templar Knights in an inner ring. An attack against his family would not be unprecedented.

"How are the girls?"

"Scared for their Papaw and father. The two oldest are old enough now to understand what is going on—at least to know that their Papaw is in danger."

"Please reassure them for me, but don't lie. Tell them I have a great duty to perform for my country, and a great duty to my men. Tell them I love them and will always love them."

"Tell *me*."

"I love you, Sarah. I always will. You were my first love, and you are where my heart has its only home."

"They're going to kill you, Michael. I've been feeling a powerful dread these past weeks. Every day you have traveled closer to the battle, to the heart of Islam's holy city, my dread has grown. I'm going to lose the rest of you."

"They've tried before—tried and failed."

"I want Maria to bring your body back to me when this is over, to be buried here at the farm. I want her to stay with me. She is my sister now. We both love you, and she has earned a place at your table."

"I'll tell her. She has always loved and cherished you."

"I won't tell you to be careful, Michael. I know that's not possible. I'll tell you to show mercy on your enemy as best you can. They're God's children too."

"As best I can, Sarah. Sarah, I want you to do something for me. At 9:30 p.m. tomorrow, I want you to play my homecoming song. Sit and listen and think of me. I'll be thinking of you and the girls at that very moment."

"I will, Michael." It was funny. Until these past few years, she had always called him Mike. Now Michael seemed to fit better.

"Then, at 10:10 tomorrow night, tune in the news channels and watch the war coverage."

"I will, Michael. I'll be watching you. Goodbye, my love."

PREPARATORY FIRES

Scofield, Maria, Mark, and Capt. Weaver were on the observation deck of Scofield's flagship when the final-phase bombing began, wearing their nano-armor to protect them from debris and enemy fire. Jeddah was already on fire and had sustained severe damage, but that was nothing compared to what the Navy and Air Force had planned.

Precisely at dusk, the air attacks intensified, with large bombs hitting almost every structure still standing. The Air Force dropped incendiary bombs around the old town and business district, igniting a ravenous conflagration.

Scofield and his comrades watched this destruction for thirty minutes before retiring to the command center to check the data feedback from intelligence-gathering devices spread throughout Jeddah and along their planned approach the next morning.

Scofield had electronic eyes everywhere on the battlefield. He could see his enemies in visible light, in the infrared spectrum at night regardless of the ambient light, and even through walls. Mouse's algorithms processed this visual data automatically, the detailed threat map composited by the computer showed that Scofield would have limited opposition as he assembled his forces in the area west of the Prince Abdullah al-Faisal Stadium and moved towards Mecca.

The first major force to oppose the Templar advance would be an elite mechanized division from Syria east-southeast of Umm As Salam, a small town outside of Jeddah. The division lay in ambush where the Alharamain Expressway turned south to cut through a few small hills. Scofield ordered the division spared during the air-attack preparation phase of the battle, so he could use this first road bump as a tune-up for the Templar division. He would order a combined air-ground simultaneous assault that would be meant to obliterate the Muslim division as they drove through it.

After passing this first test, the Templars would enter the southwestern suburbs of Bahrah, which was already being reduced to rubble for easy passage. They would then emerge on a flat plane two miles long, where they would face ten reinforced mixed divisions of the Islamic Defense Council Forces, about one hundred and fifty thousand troops, dug in along a ten-mile ridge intersecting the expressway east of Bahrah. This defensive line would be a major battle, and Scofield wanted to neutralize all of these divisions permanently.

As with the preparatory battle at Umm As Salam, Gen. Marks planned a combined-arms attack using the Templars' Navy, Marine, and Air Force aircraft—and, for the first time, the armored and cloaked Army mechanized units from base camp. The Air Force and Navy were to begin preparatory bombing at dawn as Scofield kicked off from Jeddah. The Air Force would deploy a string of MOABs (Massive Ordnance Air Burst, or "Mother of All Bombs"), dropping these massive, 21,000-pound bombs over the divisional lines of the Islamic forces on Scofield's command—another operational nightmare for Scofield's executive officer. Since these bombs are difficult to deliver on demand, Marks would give Scofield a required lag time that he would have to abide by. These massive air-burst incendiary bombs can sterilize an area several hundred meters in diameter and look like a small nuclear weapon when detonated, scaring the hell out of any survivors.

The Islamic Defense Council set up their final defensive line outside Mecca, integrating it into the western suburbs and using buildings as cover. It consisted of five mechanized divisions from five nations, each nation offering their toughest division for this final fight with the expectation that they would prevail or die. These forces would be attacked in a similar manner as at Bahrah, with the Templars breaching the center of the defensive line and plunging into Mecca to take the Holy Mosque that surrounded the Kaaba. But rather than destroy these divisions as intended at Bahrah, Scofield wanted to reach the Kaaba quickly without getting bogged down. He would use the Kaaba as leverage to force a temporary truce and establish grounds for the final confrontation outside the city. He could thus spare the city and concentrate his enemy out in the open.

Aircraft would isolate the remnants of the Mecca defensive line as they counterattacked towards the Holy Mosque, hopefully ending the first day's fighting. Scofield expected a few hundred casualties of his own, and over 200,000 from the enemy. Added to the Jeddah dead, they would bring the enemy casualties to an expected 700,000 in less than twenty-four hours.

LAST NIGHT ABOARD SHIP

"Maria, I'm surprised to find you awake," said Scofield, disrobing. "You usually sleep like a dead woman the night before battle."

"I was wondering if you had time for our pre-battle ceremony, my love."

"Not tonight, I'm afraid. We simply don't have time. If the Islamic Defense Council accepts my challenge and proposal for a truce tomorrow, we'll have time before the final battle. We should have at least two days of downtime at base camp while they assemble their forces."

"I want to make love right now, just in case."

Exhausted, and knowing he would be awakened in just a few hours, Scofield obliged, patiently and fully. He understood her fear and regretted his preoccupation. He knew how much this ritual meant to her. Looking up from her perfect body to her blood shot eyes, he saw that she had been crying. He smiled and gave her what she wanted, what she needed, that night.

LANDING FORCE

Capt. Weaver, already dressed and ready to go, woke the general at 3:00 a.m. Scofield nudged Maria, interrupting her slumber. Weaver looked deadly serious this morning—his standard demeanor when gearing up for battle. He got his charges roused and ready and accompanied them to the command center. They could hear the rumble of explosions outside the ship.

They must be close to the piers, thought Maria.

The forward elements of the Templar division were already ashore, having landed thirty minutes earlier. As expected, there was little opposition at the docks. The ships were cloaked, as were the Templar Knights, and most of the citizens and soldiers in the area were hunkered down or dead, so there were few eyes to spy them.

Before the introduction of nano-armoring, the shelling of a beach would shift to avoid friendly casualties in the first wave of a landing. In the old days, it had been obvious when a landing force arrived. With the advent of the armored suits, the shelling could be kept tight and uninterrupted. Cloaking and the close-in shelling provided perfect cover for their landing. So far, all troops had been offloaded successfully, and amphibious ships had come and gone without incident. Secrecy had prevailed.

PRINCE ABDULLAH
AL-FAISAL STADIUM

The cloaked Templar Knights assembled in a long column that snaked around an open field near the stadium. Ready and waiting for their commanders, they were invisible to anyone not connected to the global operating system (GOS). The GOS tracked each warrior, rendering a ghostlike image of him for any viewer connected to its communications web. A GOS observer could see a cloaked Templar's outline—and his face and expressions, if his face wasn't masked.

The warriors watched the intensified bombing from the cloaked aircraft flying all around them. In the days before nano-armored suits, bombs could not be brought this close to friendly forces; now they could park them within fifty yards with little fear of friendly casualties.

The city was clouded in thick smoke, broken only by the bright fires roaring out of the few remaining buildings with fuel left to burn. Periodic explosions flung rubble into the air sending up mushroom clouds of debris, smoke, and fire. Closer to their location—in the less-populated part of town—gun and troop emplacements were dug in. They could see the soldiers hunkering down as helicopters and attack planes sought them out. Explosions hit the emplacements with uncanny accuracy.

The GOS could see what the helicopters, planes, and nanowarriors saw. It analyzed the petabytes of real-time data flowing in from the multitude of cameras, thermal sights, night-vision sights, and other sources simultaneously. The vast streams of data were sent to parallel processing engines embedded in every nano-lattice frame spread throughout the command—in planes, at base camp, in nano-armored suits, and any other device connected to the GOS. This mobile set of supercomputers was able to detect enemy emplacements much better than the eyes of ordinary men, even those enhanced by Scofield's genetic elixir. Unbeknownst to the Islamic fighters, some of whom stood just yards from the Templars, the battlefield had tens of thousands of eyes, all integrated into one artificial intelligence whose sole purpose was to find them and kill them. If they had known how vulnerable they were, they would have fled in terror.

The division stood in formation, a silent witness to the bloodbath. The spectacular scene would be burned for all time into the synapses of survivors. The Templars stared in awe at the scene unfolding in front of them. Soon it would be their turn.

Gen. Michael Scofield and Col. Maria Olsen arrived in a cloaked Blackhawk helicopter near the rear of the kilometer-long formation. Dismounting the bird, they strolled along the entire lines of the Templar Division, stopping to talk with Knights, asking their names, and offering encouragement. Fed the video from Scofield's suit, each Templar Knight could participate in the ritual walk of their commander. It was a dreamlike experience, watching the general nonchalantly walking along the ranks, chatting, while all hell was breaking loose around them.

"Capt. Deissler. Glad you could make the kickoff of our little stroll to Mecca."

"Wouldn't miss it for the world, General. Would you like to meet Private Bokert, sir?"

"By all means."

Bokert snapped to attention and saluted the general.

"At ease, Bokert."

"Thank you, sir—and thank you for letting me join you here, to fight our enemy."

"Tell me if you still feel that way when we are finished," Scofield said with a wink.

Reaching the front of the formation, Scofield and Maria greeted the staff members who had been selected for the honor of accompanying them to Mecca. Most of his staff, including Brig. Gen. Marks, were at base camp, running the war from the fire-control center. The general's personal bodyguards and some command and control personnel would accompany him as the lead element for this historic march.

The division waited stoically for the time when they would cross the imaginary line of departure and the official kickoff of their march to Mecca. A cloaked Blackhawk helicopter landed in front of them, and a tiny Templar Knight jumped out. Scofield smiled as the munchkin in her small, armored suit approached. Carrying his second Medal of Honor (he was wearing the first under his nano-armor), she stopped in front of her general, nodded her head in fealty, and waited for him to kneel before placing the medal over his neck. She had cloaked and armored the medal to make sure it would blend into his suit but be seen by all Templar Knights. She kissed him on his bowed head and then returned to the helicopter. As she boarded the bird of prey, there was a thunderous roar of approval from the Templar Division.

At the appointed time, the general gave his commander's last-minute instructions and encouragement and ordered the division to move forward. The long formation crossed the field to the edge of Falah Street, turning east to Mecca. As they entered the street, each Templar row turned off it's cloaking.

The division was slowly unveiled, rank-by-rank. Few enemy fighters, dying or hiding, saw this amazing effect, but the world would see.

The moment that Scofield crossed the line of departure, the entire world television audience was instantaneously transported to the battlefield at Prince Abdullah al-Faisal Stadium. Until that second, Mouse had fed pictures of the Jeddah battle from the cameras she controlled on the amphibious ships. These ship-borne images were as dramatic as any real-time combat footage had ever been, and most networks aired Mouse's video as their primary feed.

Suddenly, the video switched from remote Jeddah images to a shot of the Templars emerging from hiding, deep within the battlefield itself. Mouse didn't explain her sleight of hand, leaving it to the news anchors to decipher what they were seeing and where it was coming from.

This visual trick was accomplished with another bit of nano-enchantment from Mouse's magic box: images sent from flying cameras. The battlefield was littered with cloaked floating recording devices. Designed for intelligence gathering, they made for superb television, maneuvering like sport cameras over a football field—independent of wires. The visual effect was the same: a super-stable, highly mobile, high-definition picture from the point-of-view of the individual soldier.

The viewing audience was awestruck by the grandeur of the suited warriors emerging row-by-row from hiding, and the outward zoom that slowly revealed scenes of the battlefield. They gasped at the panoramic view of Jeddah, with its wide destruction and the bombs raining down.

As the Templar Division turned east on a wide road, a computer-generated image of the street's name [FALAH STREET▶] was injected into the live video feed. Two three-dimensional road signs also materialized in the video feed: [◀JEDDAH, TWO MILES], and [MECCA, THIRTY MILES▶].

THE MARCH FROM JEDDAH

The lead elements of the division quickly traveled the five miles between them and the first Muslim division lying in ambush along the route at Umm As Salam. They encountered sporadic contact with enemy fighters along the way, but not enough to warrant cloaking or to impede their progress. Though their location was broadcast to the world, it filtered down slowly to the ground commanders, who were busy surviving the constant bombing runs against their positions.

The Muslim commander of the Umm As Salam ambush could see the lead elements of the division approaching his lines. He'd organized his defense in a classic "L" ambush, with a blocking force facing the approaching Templars perpendicular to the Alharamain Expressway, and a larger force set back about 400 yards, parallel to the road. He held fire, waiting for the formation to get closer and deeper into his kill zone. He would draw the Templars toward his blocking force and have the northern arm of the L open fire when they were strung out and engaged. His troops were dug in, well supplied, and supremely motivated, worthy adversaries for this apostate general.

It was a good plan, except that Scofield was privy to every detail, down to the disposition of each soldier, gun emplacement, and tank. Scofield was watching a video from a flying camera positioned behind the Muslim general. When the general raised his arm to ready the first phase of the attack, Scofield pre-empted him.

Gen. Marks at base camp had the same intelligence as Scofield, and he'd planned attacks against every gun and tank position from helicopters and low-level fighters flying over the unsuspecting Umm As Salam defenders. Scofield had assigned a company of Templars to each enemy regiment, with detailed intelligence and plans for killing them. At Scofield's command, the Templars cloaked and scattered as Marks' aircraft fired their weapons.

The live video feed supplied to the news organizations switched to the spy camera behind the Muslim commander, so that viewers around the world could see what the Muslim general saw. They bore witness to the long Templar line marching towards them down the road, the Muslim general raising his arm to signal attack, the Templars vanishing, and the thunderous explosions along the general's L ambush.

The perspective then shifted to an angle forward of the Muslim general's command bunker, allowing viewers to see him die in a great explosion, sending a mushroom cloud of debris high into the air. The camera then flew along the lines of the ambush, showing men firing in all directions at unseen enemy tormentors and being slaughtered en masse.

The battle was over in thirty minutes. That was the time it took for a Templar Battalion, backed up by nanoclad aircraft, to completely destroy a conventional mechanized division. Not a single Muslim soul survived.

The camera panned the dead soldiers and then zoomed back to the Templars uncloaking and reforming their line to resume their march. The electronic sign now read [Mecca 22 Miles ▶].

THE BATTLE OF BAHRAH

The Templars force marched three miles to the edge of a hilly area just west of the city of Bahrah, pausing there to receive their final mission briefings. The city of Bahrah and the Bahrah defensive line to the east had been pounded by air attacks all night and throughout the morning. The ten divisions facing the Templars were dug in along a series of hills three miles to the east of Bahrah.

This defensive line was especially dangerous because of the concentration of large weapons that could defeat the Templar armor. Scofield knew that the Muslim commander wasn't planning an ambush, hoping instead that the Templars would run along the road in the same arrogant fashion as they had done at Umm As Salam an hour earlier. He intended to open fire as soon as the first rank showed itself, and had mined the road and planned artillery fire at obvious approaches to his line.

Scofield turned off cameras to the outside world, then cloaked his division before cresting the last hill to spill out onto the plane south of Bahrah. While the Templar division deployed for their frontal attack, he called in the MOAB attack to hit the entire Bahrah defensive line.

To the uninitiated, a MOAB explosion looks like a small nuclear explosion with its large mushroom cloud towering hundreds of feet high. To the Muslim fighters on the ground, it was hell on earth. The killing potential of this Templar force was being fully revealed to the enemy, shattering the resolve of the ten divisions under attack.

There was a lull in the battle as the Templar warriors began their long approach to the Muslim defensive line and the air attack planners waited for the dust to clear from the MOAB detonations. As soon as Gen. Marks could assess the damage, he committed his attack aircraft to the battle to destroy the remaining gun and tank emplacements. Marks' air assault continued for twenty minutes while Scofield's Templars closed the distance to the defensive line.

For the cloaked Templar attackers, the Bahrah defensive line was anti-climatic. They were left with the mop-up job of killing the men and destroying the equipment of remnant units terrified, disorganized, and disoriented. The Templars went about their gruesome task with deadly efficiency.

The battle of Bahrah was over in three hours. It was impossible to kill all of the one hundred and fifty thousand Muslim soldiers without chasing them in a thousand directions. The estimate of the dead exceeded one hundred thousand. Any wounded not slaughtered by Templars were left to die. There would be no prisoners.

THE FIRST BATTLE OF MECCA

The Battle of Bahrah positioned the Templar Division eighteen miles from its final objective: the Holy Mosque. The Templar forces had to face five more divisions dug into the outer suburbs of Mecca.

Keeping his division cloaked, Scofield moved as fast as he could to reach the outskirts of Mecca, airlifting one battalion to the rear of the Mecca defensive line to shorten their combat time. The Second Battalion First Templar Regiment would spearhead the Mecca battle, being ferried over the course of an hour from the Bahrah battlefield to Mecca in multiple waves of cloaked helicopters.

While the Second Battalion was positioning to Mecca, Scofield started his Templar Division on a grueling forced march along the Alharamain Expressway. The usual pace of forced march for a Templar was a nine-minute mile. That pace would put the division in the attack in a little less than three hours. Scofield increased the rate of march to six minutes per mile, which would put them in battle an hour earlier. A six-minute mile (a dead run) for eighteen miles in the desert heat would kill most combat-loaded Marines, even those in the best condition. With nanocooling and nanomuscular support, it was hard but not deadly. The division buckled down to their task and moved to contact in good order.

At the one-hour mark, the Second Battalion began their attack from the rear of the center of the Muslim line. Gen. Marks coordinated the air attacks, concentrating them on the center just in front of the Second Battalion as they attacked and along the edges of the battalion's advance, to ward off reinforcements from the enemy on either flank. The attack by the Second Battalion and Marks' aircraft cut through the Muslim center ten minutes before Scofield reached the line.

The pace of the Templar attack was so quick and brutal that the Muslim divisional commanders were unaware that their line had been breached and that the Templars were pouring through. Scofield rushed his division into the heart of Mecca and surrounded the Holy Mosque before his enemies could effectively react. The Templar Division created a defensive position one mile in diameter around the Mosque. They would have to repel counterattacks against the outer edges of their perimeter, but the Muslims didn't dare attack near the Holy Mosque or the Kaaba.

Nano-armored warriors were most effective when they were cloaked and on the move. In a static position such as they found themselves at the Holy Mosque, fighting massed tank and artillery fire, they were vulnerable. The division started to take casualties as they held their perimeter and awaited their general's instructions.

When a Templar Knight was hit by a penetrating round, the result was horrific. The energy of the round penetrated the armored suit but rarely exited, creating an implosive effect that literally blew the warrior to pieces. Typically, the suit would lose cloaking, and the implosion could be recognized by the Muslim forces as a "hit." Sometimes the nanoclad warrior would be hit in a leg or arm and lose the limb violently.

It was during this phase of the war that the Muslim commanders learned of the limitations of the nano-armor suits. They ordered all available large-caliber weapons with armor-piecing shells to be brought to bear against the Templars surrounding the Holy Mosque.

THE TRUCE

A flying camera was positioned north of the Kaaba. There were no people in view, but the television audience could hear the rumblings of explosions, aircraft, and other commotion. Lacking input from the Templar staff, the television anchors told their viewers the significance of the Kaaba, Islam's holiest relic and site, explaining that Muslims pray five times a day, turning toward Mecca and the Kaaba to pay homage.

Into this scene strode Gen. Michael Scofield, towing a cart with a cone-shaped warhead. "My name is Gen. Michael Scofield, United States Marine Corps. I am standing near the Holy Mosque and the sacred place of pilgrimage of the Kaaba. The United States forces will agree to leave this holy place unharmed on two conditions: First, that all Islamic forces stand down their attack immediately and agree to a two-day truce while our forces reposition to a battlefield east of Mecca. Second, that Sheik Abdullah Caliph al-Rashid present himself to me, together with his champion, to engage in personal combat to the death.

"If you don't agree immediately to this truce and all of my terms unconditionally, I'll withdraw my forces and detonate this ten-megaton bomb two feet from the Kaaba. I await your response. One of my Templar Knights is standing in your command bunker. You may notify him of your decision."

Scofield could see the shocked faces of the assembled Muslim leaders in the bunker of the Islamic Defense Council through Capt. Bob Mopar's helmet cameras as he de-cloaked before them. He heard, though Mopar's translator, the mumbled consent of their leaders, and their frantic attempts to contact their forces and have them stand down.

"I have recorded your consent. As soon as I hear that your forces have disengaged from mine, I'll withdraw to our new assembly area and remove this weapon from the battlefield. As promised earlier, I'm now granting reprieve to the one thousand terrorists on the list handed to Sheik Abdullah Caliph al-

Rashid to travel to this location. If any of these cowardly men don't show up to fight, we'll return here and conclude our business."

On cue, Mouse shut off the camera feeds, replacing them with roving views of the earlier battlefields at Jeddah, Umm As Salam, Bahrah, and Mecca: wrecked equipment, mutilated bodies, and burning buildings.

ASH SHARĀ'I`

"We'll meet the enemy on the planes west of Ash Sharā'i`," Gen. Scofield addressed his senior commanders in the base camp briefing room. "Our forces will be situated south of Highway Forty, facing north toward the chosen thousand, four hundred meters off. The rest of their armies shall be arrayed any way they wish, as long as it's beyond a one-thousand-meter buffer. Once the battle begins, there are no rules," he paused, "of any kind."

Scofield drew an arc on the digital map showing the allowed enemy dispositions around his intended formation. The resultant image looked like a fat vase, the top consisting of the line facing the Templar Division, and the rest of the vase outlined by the surrounding Muslim divisions. The Templar formation looked like a small block of wood inside the vase. If the men watching this brief hadn't just defeated sixteen Muslim divisions, they might have had misgivings. Scofield was willfully allowing himself to be surrounded by a vast army.

"Inform our enemies that they must be ready to meet us at 9:00 a.m. two days from now. Let me know when all the arrangements are complete.

"Move the division to the assembly area immediately. We must remain vigilant against subterfuge as we rest and wait for the next two days. The GOS and the cameras should alert us to any threats, but we must remember that two Iranian nukes are still unaccounted for, and we cannot exclude the possibility of outside intervention—from Russia, China, or India. Our strongest deterrent against a nuclear attack at Ash Sharā'i` is the certainty of our own nuclear counterattack.

"Rest the division as best you can. I'm afraid the flyboys will have to work around the clock. I'll remain at the base camp and hold daily briefings. Admiral, what is the status of the fleet?"

"We own the Red Sea operational area, on the surface and below. We also own the skies in a four-hundred-mile radius around the fleet and the Mecca area of operations,"

"Good. What about our Iranian friends? Do we have new intelligence? Col. Franks?"

"We're making progress, General, but we haven't located the main conspirators. I'm afraid the world is wising up to our cloaked spymasters and taking countermeasures. We've apprehended fourteen of the targeted sixty-five leaders, but unless we have a breakthrough, we may not find the rest until after the battle."

RATTING OUT THE RATS

Col. Melissa Franks stopped by Gen. Scofield's quarters to announce a breakthrough in the Iranian roundup. An anonymous message had come in to the communications center on an encrypted link detailing the hiding places of all fifty-one Iranian leaders yet to be found. Wary of a trap, she had proceeded cautiously, but the first two locations had proved genuine and they had apprehended the targets. With that, she moved fast, snaring the rest before they could discover their vulnerability. All sixty-five Iranian political and military personnel on her list were now in captivity and being transported to base camp.

"Have you determined the source of the information?" Scofield asked.

"Not yet. The tipster disguised his identity. We believe it came from our link with the Islamic Defense Council."

"Thank you, Melissa. You do good work. When do these murderers arrive?"

"Within the hour."

"Set up a place for the interrogations and executions."

"Yes, sir. May I do the honors?"

"I'll take care of the President. You can divide up the rest."

Listening to the exchange between Scofield and Franks, Maria looked troubled. "Did he arrange it?" she asked.

"Likely."

DEBT PAYMENT

Melissa Franks met Gen. Scofield as he entered the enclosure housing the captured Iranians. "General, we've interrogated the Iranians," she began. "They've admitted that the last two bombs are in separate locations. One is in the hands of Gen. Avizhed, who commands the Iranian Revolutionary Guard. The Guard has been left in Iran for internal security reasons, and the exact bomb location is unknown. The Guard plans to deliver this bomb during the battle.

"The other bomb was reported to be in Chicago, set to go off tomorrow before our little get-together. The general who gave us this information doesn't know the location of the Chicago bomb. We've alerted the FBI, CIA and Homeland Security. I've also vectored two squads of Templar knights from our US reserves to Chicago in case stealth is required. They'll keep us informed. President Donner was informed of this information and has convened a cabinet meeting to discuss what to do," Franks concluded.

"Are they going to evacuate Chicago?" the general asked.

"Don't know, but I don't see how they can do that and not alert the terrorists. I would think they would try to find the bomb first," Franks said.

"Yeah, makes sense. That's a tough call to make," Scofield agreed. "Anything else before we start?"

"I promised the general who gave me the information about the bombs that I would spare his life. It's probable, however, since he had detailed information, that he was directly involved in the first two explosions."

"*You* promised him his life. *I* did not. Bring him to me."

The man was brought to Scofield, looking like he was about to face the devil himself. "You have told the whole truth? Is there anything else you want to confess?"

The man shook his head.

"Were you involved in the first two explosions?"

Again he shook his head.

"How is it possible that you know about the Chicago bomb in such detail, but not the New York or Washington bombs?"

He shook his head again.

"You are lying," Scofield glared at the Iranian general trying to discern his purpose, then slowly unsheathed his sword and behead him. "You lying murderous bastard."

He then walked slowly to the line of Iranians and swiftly lopped off the head of the Iranian president. He then paused and nodded to Franks and Pvt. Bokert to finish the job.

"But…" Col. Franks started to question the wisdom of this but Scofield signaled curtly to get on with it. He stood stoically and watched the proceedings.

"Col. Franks!"

"Yes Sir."

"Find those fucking bombs!"

"Yes Sir." *You might have wanted to wait a bit before beheading these hagis.* Melissa Franks thought. *Now what am I suppose to do?*

DINNER WITH FRIENDS

"It seems like eons since we last sat down for a feast together," Scofield lamented. "Let's do justice to this meal. I know it isn't much, but we eat what the troops eat. They're in the field, and we're here in comfort, so we should be grateful."

"But I'm a civilian, and I'm starved," Mary Louise teased.

"I could rectify that if you wish, Private Monahan."

"Uh, that's okay. I'll be fine with these table scraps."

While Mouse and the general bantered, Maria got a head start, hiding their food under a nanoblanket as she devoured her own.

"You are not too old for me to spank, young lady."

"My plan all along," said Maria, smiling as she returned their rations. They were tasty enough, with over 8,000 calories in the two food tubes. Hers were banana and peach flavored after switching with Scofield. When the general saw he now held coconut and spaghetti he frowned his disapproval.

They ate in silence. No one wanted to talk about the horrors they had witnessed, or the dangers faced. This was relaxation time, Mental Health 101 for warriors needing to decompress.

Of course, no meal with Scofield was complete without his after-meal ritual, but that wasn't what made them laugh this time. Watching him strain was hilarious. The tube rations had the aftereffect of constipating the consumer— and he'd had three such meals that day. He wouldn't give up, though, and, to the chagrin of all in attendance save Scofield, eventually delivered his report.

In the silence that followed, the three friends sat peacefully, delightfully drained, absorbed in their own private thoughts. To interrupt this perfect serenity would be an offense. Quiet was good—a gift they could enjoy together.

THE CHICAGO BOMB

It was late when Col. Melissa Franks knocked on the general's bedroom door. Seeing Maria and Mouse in bed with Scofield, she felt a twinge of guilt for disturbing their peaceful solitude. She wasn't without envy and heartsick that her insanely brave husband, who might be killed in the forthcoming battle, could not share her bed that night. She shook off her melancholy and woke Scofield gently.

"We think we found the Chicago bomb, General. But there is a problem," she whispered excitedly, trying not to wake his bedmates."

"What's the problem, Colonel?"

"It's armed and well guarded. The FBI has operational control but they won't let our knights help them. Fucking turf war."

They both stepped outside the general's private enclosure to speak.

"Do the bombers know we've surrounded them?"

"Nobody knows. The FBI guys are considering talking to them. I guess they think they can be reasoned with. They have some Ph.D. level negotiators who are saying that, ah, deep down, everybody wants to live…"

The general nearly exploded at hearing this. "Are you fucking kidding me, Colonel? These guys? Really?" He was finally at a loss for words.

"I'm just the messenger, General."

"I'm sorry Melissa, yeah, I know. What do you recommend we do?"

"We can cloak some Templar Knights and take charge," Melissa posited.

"Call Donner and relay our conversation. Even if he doesn't give us operational control, he can stop the FBI guys from committing suicide—for the whole city."

PREPARATIONS FOR THE FINAL BATTLE

The day before the battle at Ash Sharā'i` was a busy one. Scofield and his staff spent most of the day watching enemy troop deployments via the GOS spy systems. It was amazing watching an army of over two million men assemble for battle. Scofield was glad that he had given himself two days to prepare, so that the Muslims had time to bring all their war-fighting equipment to the battlefield. It also gave him time to stockpile munitions and plan the air targets.

Scofield knew that his Templars could not possibly kill that many men and destroy that much materiel. The logistics just didn't add up. The Templars were a focal point, a means of drawing the enemy in. They would do their part. They would have a significant psychological impact, and kill thousands. Nonetheless, the majority of the killing would come from the air—airspace that his forces owned with absolute domination.

At great length they pondered the psychological warfare (PSYOP) plan. Scofield wanted to break the fighting spirit of the Muslim forces early, making them appear weak to his viewers while he hunted them down and killed them. The day grew long and tiresome with the many details, but Scofield wanted to sleep before this battle—and leave time for his devotional ritual with Maria, which might be their last.

At 10:00 p.m. Ash Sharā'i` time, he excused himself, telling Weaver to wake him at 4:00 a.m., but make sure that no one disturbed them again until 8:00 a.m. Weaver understood why without asking.

FAREWELL, MY LOVE

Over the years, Scofield and Maria had developed a private ritual to reaffirm their devotion to each other prior to battle. Each movement was choreographed to demonstrate their reverence for each other, the ceremony culminating in a profound sexual coupling intensified by the deep emotions they expressed and the imminence of combat. Maria had invented the ritual, leading Scofield like a partner in a waltz. It was their shared masterpiece.

The unspoken realization that this would be their last communion added poignancy as the time of departure approached. The ceremony could not be delayed much longer, but both were reluctant to begin, knowing that it would have to end.

Maria disrobed to bathe and beautify herself. Her preening was the prelude, but the general, would initiate the ritual when he was ready, not before. As she prepared, Maria became preternaturally silent, performing her tasks soundlessly, with hypnotic effect. She relaxed completely, her mind focused, trance-like, on Scofield. She would not take her eyes off him until they were finished. Scofield waited until the last moment to acknowledge her gaze, signaling his acceptance and his reverence for her gift of devotion.

The general began the ritual by disrobing and anointing himself with oils. Standing with his legs wide apart and the oils running down his body, the general spread his arms outward, palms raised to the ceiling. He lifted his eyes and waited for Maria.

Maria approached silently, touching Scofield softly, sending shivers down his spine. Reverently, she rubbed the oils into every inch of his skin. She anointed herself by rubbing her own body against his, moving slowly and methodically, putting them both into a dreamlike state.

Maria embraced Scofield, hugging him tightly until he enveloped her in his powerful arms. There they remained, lost in their own thoughts. The ritual had its own pace, which neither of them would break, but this time was different, and Maria began to cry.

She slowly rose to kiss his lips. She was ready to give her heart to him one last time.

As Scofield lifted her up, she wrapped her legs around him. They stood there motionless, raising each other to the edge of climax with their minds, feeling their physical connection, but resisting the desire to stoke their passion.

They finished silently together, holding their embrace to let the passion drain from their bodies.

Time passed slowly as each held their embrace. Finally, reluctantly, Scofield glanced at the clock on the wall. He knew that they could delay no longer.

"Are you apprehensive, my love?" Maria asked.

"I've already died once; there is no mystery for me. When I was brought back, I was angry that I'd been taken from the warmth and beauty that had enveloped me. I'm ready now as I was then, yearning even. The pain of dying is a small price to pay for the beauty that awaits."

"I'll miss you."

"And I you, my love, but we'll find each other again. We'll join our voices with the angels and spend eternity with Sarah and the girls."

THE NAME OF THE BEAST

As Col. Olsen and Gen. Scofield left their enclosure at base camp to hop a ride to the battlefield, they saw Mouse hurrying toward them.

"I know who he is, General," Mary Louise said excitedly.

"What is your confidence level?"

"Very, very high."

"Is he here?"

"We don't know. He is a second-tier Saudi Prince. He fits every profile."

"Put his picture on the GOS to search for his face. If you locate him on the battlefield, order our forces to protect him, and then call me, I'll deal with him personally."

THE PARADE OF DEATH

The Templar Division was organized in companies, each company forming a near perfect square of two hundred and twenty five warriors standing exactly two feet apart. The companies formed battalions by positioning three companies in offset phalanxes to the front, with one company to the rear. The battalions formed regiments in the same manner, with three battalions to the front and one in reserve. The division stood the ground with three regiments on line.

The sight of the Templar division arrayed in the classical formation of a Roman legion was striking. Tightly grouped formations like this were not practical in modern warfare with modern weapons of concentrated lethality, but the spectacle was a grand one. Flying over the division, cameras showed the Templars standing toe-to-toe like black chrome statues, appearing utterly unafraid. The flying cameras also showed the spectacle of the massed Islamic armor, men, and cannon in a seemingly chaotic array, highlighting the dichotomy between the two forces, the vast technical gulf that separated them, and the great disparity in their numbers.

Marching as a unit in perfect cadence, the Templars moved over the hill to a position on the down slope where the forward elements of the Islamic formations could see the entire division.

Both sides went quiet as the Templars stopped their advance. The battle would not begin until the confrontation between Abdullah Caliph al-Rashid and Scofield was completed. There was time to admire the raw beauty of this spectacle.

A lone Templar Knight, a female uniquely adorned with a scarlet-plumed helmet, stepped forward from the lead company in the center of the formation. Maria had chosen this flourish to draw attention to herself while she sang a song for the dead—a feminine look to provoke a maternal response from her Templars, and to rile the Islamic chauvinists.

She began to sing, her amplified voice filling the valley with beautiful music which evoked a profound sense of sadness and resignation in all who heard her, Templar and Muslim alike. Mouse captured the audio for the television audience, and they too were moved by the beauty of Maria's voice and song, which rose and fell in a haunting melody only one other listener in the world could recognize.

The Islamic army could see the image of Maria's face displayed on the suits of the Templar division, each Knight a part of a giant pixilated screen, and some of the Muslim fighters recognized the internationally known celebrity, the singer/warrior fighting for her country.

"General, what is that song? And what language is Maria singing?" Mouse asked Scofield over their private communication link.

"It's the most ancient language, the first language. Only the angels speak it. The song is straight from heaven. It's a song of prayer for the dead, calling the angels to this place to care for the dead."

"Where did she learn it?"

"She once told me that it was given to her by angels, along with her voice, to soothe her sorrow and comfort the dead when we viewed the corpses laid out from the Third Regiment after our defeat in Afghanistan."

"It's so beautiful."

"Yes."

Maria's lamentation came to a mournful close. She had given every ounce of her being to the song and felt drained. All could see the look of rapture on her face.

When she finished, Gen. Scofield rode out from the formation on a red warhorse, his sword laid casually across his saddle. Even for a Friesian, originally bred for war, this horse was large and powerfully built. As he left the Templar formation, Scofield transitioned his suit from jet black to the image of a medieval Templar Knight in armor, with the sign of a red cross emblazoned across a white background. He trotted to the center of the valley between the opposing forces and stopped to await his adversary. New music filled the air with anticipation and dread.

Sheik Abdullah Caliph al-Rashid and his champion walked out toward Scofield. Abdullah had replaced Osama bin Laden after his death as the symbolic leader of Al Qaeda. A figurehead like bin Laden, he was revered by the fighters who stood behind him. His champion was almost seven feet tall, powerfully muscled, and armed with a large Arabian double-edged broadsword. The sheik strode purposefully to his fate, expecting to be martyred. He had prepared his body in the ritual way, praying to Allah to fortify his resolve in accepting the gift of martyrdom. His murder would inspire his people and lead them to victory. He would die passively after his champion was vanquished, robbing the apostate usurper of any satisfaction. His reward would come in heaven.

The two adversaries met, the sheik's champion standing behind him, sword at the ready. Abdullah Caliph al-Rashid spoke first, in Arabic, his words automatically translated into the languages of the world: "You ride here, usurper, wearing the color of the horseman and of the ancient crusaders. You are an apostate, a liar, and blasphemer. You'll burn in the fires of hell."

"Silence, al-Rashid! I've no interest in your words. I've come for payment of a debt, and your death is the down payment. I'll chase your soul to the gates of hell." Scofield moved past the sheik to confront his champion, lightly pushing him aside. Abdullah Caliph al-Rashid lost his balance and fell to his knees. Scofield's indifference to him was a great insult.

Scofield addressed al-Rashid's Afghan champion directly in Pashto: "I honor your courage for facing me this day, warrior. Make peace with your god. You are about to die."

The huge man scowled. "Prepare to die, infidel." He lifted his large sword and swung it powerfully towards Scofield. The general easily parried the blow. Surprised at the old man's agility and speed, the champion raised his sword for a downward thrust that would drive the infidel into the ground, even if he blocked the heavy sword with his own. Out of the corner of his eye, he saw the infidel's sword glow bright blue.

With a deft flick, Scofield cut the champion's sword off at the hilt. Before he could respond, Scofield swiftly severed the man's hands at the wrists. The champion gaped in horror at his stumps. Scofield flicked his sword twice more, severing the man's legs below his hips. The general rushed to the man to steady him at his shoulder and jam his stumps into the ground to preserve his consciousness a few moments longer. The champion, now in shock, looked at Scofield in terror and helplessness, pleading with his eyes for mercy. Scofield then placed the edge of his sword atop the man's head and slowly sliced through his head and body, cutting him in two.

Seized by fear, Abdullah Caliph al-Rashid staggered backward. Scofield stared at the frightened man, hoping he would run. When he hesitated and looked as if he might summon his courage, Scofield took a step towards him. This was enough for al-Rashid. He fled like a man chased by demons, crying and calling for his men to help him. His panicked words were translated for the worldwide audience as he stumbled, crawled, and scurried away from Scofield. His army witnessed the ordeal on the Templar display.

Locking his eyes on al-Rashid, Gen. Scofield pulled a device from his pouch and released it into the air. The nanoseeker net unfolded and flew toward al-Rashid, propelling itself to its target with hundreds of tiny nanojets on its tendrils. Al-Rashid fell to the ground, screaming in fear. The net enveloped him, holding him in a vice grip while Scofield slowly approached the humiliated man. This man had given final approval to the nuclear attack on America and had bragged about it on a video released to rally his forces to this battlefield.

Loyal to the sheik, the Islamic soldiers watched his capture and disgrace. They prayed that al-Rashid would regain his courage and die bravely. Scofield reached al-Rashid and matter-of-factly, without words or hesitation, sliced off his hands. The sheik screamed in pain and terror, but stopped suddenly as he felt Scofield's sword against his neck. Scofield slowly cut off the head of al-Rashid, maximizing his pain and suffering.

Scofield placed Abdullah Caliph al-Rashid's head and hands into a nano-armored bag and turned on its cloaking modality, handing it to a cloaked Templar sergeant who had been standing next to him.

Bullets started to ping off the general's suit. Scofield looked up to see the whole of the Islamic line turn towards him—and one man in particular glowering—a young man he had fought in Afghanistan. He ordered his nanonet to release the dead body of the sheik and "pursue and kill at slow pace" the object of his stare. As the net lifted and sought out this enemy, he fled into the ranks of the Islamic fighters. The net pursued and encircled him, slowly constricting, cutting him into a thousand small cubes of flesh, bone, and muscle.

As bullets rained toward Scofield, and larger, more-threatening guns turned to target him, he spoke into the communications channel accessing his whole command, and forwarded to the world: "Kill them!" he ordered. "Kill them all."

THE BLOODIEST BATTLE EVER

The moment Scofield said those fateful words, the black suits of the Templar Division assumed the color of blood, each warrior's chest displaying the crimson Templar Cross against a pure white background. Their formations shifted to an offset pattern, allowing the two front rows to fire their chemical bullets at the massed Muslim army. The simultaneous impact of many thousands of chemical bullets hitting the front ranks of the Islamic formation was devastating. The first two rows vaporized, bloody limbs flying into the air and enemy fire suspended as new fighters worked their way to the front. Unable to comprehend what had happened, they moved forward, meeting volley after volley of deadly accurate chem bullets from the Templars.

As the withering fire was decimating the Islamic front lines, an unseen mass of Blackhawk, Cobra, and Apache helicopters hovering just over the Templar Division uncloaked and fired a barrage of antitank and antipersonnel weapons at the front-line units. The ground heaved with the impact of their weapons, obliterating the frontline armored units before they could target the Templars.

Having achieved their psychological effect by disclosing their number and location, the aircraft re-cloaked and moved forward, increasing elevation to try to avoid the blind targeting of ground-to-air weapons. As the forward Templar units assumed a line formation, doubling their length, the GOS system under the command of Gen. Scofield ordered an attack of deadly sound energy against the Muslim fighters. As the energy reached killing frequency, driving loose sand and gravel before it, soldiers by the thousands fell to the ground clutching their ears and heads. With their eardrums bursting and inner ears collapsing, they were unable to maintain their balance and crumpled to the earth in excruciating pain. Most would die a horrid death from brain injuries and shock trauma.

Scofield ordered his reserve units in the Templar rear to cloak, then released them to their commanders for marauding. In company units, the Templars maneuvered to key areas of the battlefield, targeting headquarters and command and control centers, killing as many of the enemy as possible along their path.

With the GOS, there was little chance of friendly aircraft, artillery, or other weapons inadvertently targeting the cloaked Templar forces as they mingled with the enemy. Because the GOS knew where every nanolattice-controlled device was located and didn't allow other nanocontrolled weapons to target friendly forces, the men controlling the killing machines could fire at will without worrying about injuring their own, vastly increasing their lethality and war-fighting effectiveness.

Despite the efficacy of the Templar attacks against them, the Muslims managed to kill many Templar Knights by firing volleys of large caliber anti-armor and thermal weapons into their formations.

Scofield received instantaneous updates on friendly casualties, each Templar suit maintaining a running health check on the Knight it served. When a suit exploded violently enough to disrupt its programs, the Knight was listed on Scofield's missing-in-action tally, his status quickly updated to killed-in-action when the program from an adjacent suit automatically reported the event to the GOS. Wounded men were listed in a separate tally.

Scofield stood at the front of the formation, managing the battle with his mind, voice, eyes, and hands—like an agile conductor directing a complex orchestral work. His suit was finely tuned to his every nuanced command. Viewing the entire battlefield on his GOS-fed three-dimensional screen and ignoring the external commotion around him, he conducted his masterpiece of death and destruction.

Scofield kept an eye on his casualties while directing weapons and troop resources on the battlefield. Leaving the majority of the Templar Division uncloaked was part of the PSYOP plan. If he could panic the opposing forces into fleeing—the Templars were extraordinarily frightening—the battle would effectively be over, the slaughter would become one-sided, and far fewer Knights would be killed.

But the Muslim armies were not cooperating. They didn't seem ready to panic, and he was losing too many Knights. Scofield's battlefield intelligence indicated that the Muslim lines were not breaking because the rear formations didn't understand what was going on up front. Scofield had been too effective in crushing the three C's: Command, Control, and Communications.

It was at this moment that the Islamic generals attacked with their suicide planes. The aircraft were VTOL (vertical takeoff and landing) "jump jets" that had been brought undercover to the front lines of the battlefield. The cacophony of the battle hid their takeoff plumes and their proximity to the front lines made their delivery almost instantaneous. Ten jets, packed with high explosives and shrapnel landed amidst the Templar formations almost simultaneously.

The explosions were deafening, they sent large plumes of smoke and debris into the air. When the commotion subsided Scofield could see the damage that his computers were relaying in numbers. Parts of three battalions had been wiped out, over nine hundred dead in the first wave.

"Holy shit, Marks! How the hell did that happen?" Scofield connected to Gen. Marks on the open command channel.

"Don't know, Lion—they came from nowhere." Marks paused, then Scofield overheard him cursing to someone else. "Get them outta there, damn it." As Marks said this, Scofield saw a massive number of rocket trails headed overhead. His enemy was filling the sky with surface-to-air missiles in hopes of hitting Scofield's massed air assets.

Explosions wracked the sky as missiles hit planes and helicopters flying overhead. The battlefield became their graveyard as they slammed into the earth.

"Elevate and randomize the flight patterns of the aircraft but they must stay on station, they are critical to our success. Assign some to interdict the suicide jets," Scofield said calmly to his harried general.

"We just lost twenty-six birds. You won't have an army if we lose aircraft at this rate," Marks said angrily. "Cloak the division and scatter them into the enemy lines to disrupt their attacks."

"We need to give the PSYOP campaign time to work. Buy me some more time," Scofield said.

"Goddammit, Mike, fuck your PSYOP campaign! Men are dying out there," Marks said. As if to accentuate his point, five more suicide jets rammed into the Templar formations. Again the smoke and debris hid the damage relayed by the computers—two hundred more dead.

"Message to all regimental commanders, cloak the division and move away from the formation as fast as you can," Scofield commanded.

As the suicide explosions registered with the survivors on the Islamic front—which gave them some respite from the battle—and with the Templars cloaking in response, it gave rise to a general cheer amongst the Islamic forces. Cheers spread and seem to be coming from everywhere at once.

Oblivious to his own safety, Scofield was exposed in his role as conductor of the battle. He could have remained in the rear where he could be protected, but he always led from the front. To compensate, Capt. Weaver had asked Mouse to construct a handheld nanoshield that Weaver and his bodyguards could use to intersect incoming rounds targeted to Scofield. Forming a circle around him, they had been hit numerous times. With more rounds impacting near Scofield as the battle progressed, Weaver became alarmed. He was greatly relieved when Scofield ordered the division to cloak, and physically forced the general to a position one hundred meters closer to the Muslim lines. Deep in concentration, Scofield scowled as he was pushed forward, then smiled contritely at his beleaguered chief bodyguard.

The pace of the battle re-intensified as the division cloaked and moved forward. If the lack of feedback from the frontline was limiting the psychological impact of the battle on the rear echelon Islamic units, Scofield intended to bring the battle to them. *This will be their last cheer,* he thought to himself as he contemplated his losses.

"Gen. Scofield, Adm. Franklin here, we've located the origin of the VTOL jet launches and I've vectored Navy attack jets to their location. Now that you have cloaked we should have time to kill them. I'll let you know when we think we've eliminated that threat. Also, we've had similar suicide attacks along our entire supply line from the Red Sea to our base of operations. We can sustain operations with current losses but we'll need to be vigilant."

Franklin continued his update without pause: "We've also attacked their ground-to-air assets which have gone silent for now, they've either been killed or they've exhausted their missile resources."

"Air superiority, Admiral?" Scofield could not let it lie.

"My fault, Lion, my fault. We didn't see their movement nor did I anticipate this strategy. My bust."

"Okay, okay. Let's move on and get back on track. Do you have enough air assets to continue the ground support attacks while still killing their last air assets?" Scofield asked.

"Yes."

Scofield directed the remaining Air Force and Navy aircraft to drop their weapons beyond the frontlines—to stress the second echelon commands, rattle the less-experienced troops, and isolate the main battlefield from reinforcements, and, hopefully, start a general panic.

Scofield could see the air assets in his virtual depiction of the battlefield. Unlike commanders in the old days, who would hop into a helicopter to get such a vantage point, he could point his finger at an icon and get this greatly enhanced view. Combining the air-asset display with threat and friendly-force displays, he had a complete view of the battle space and could direct air assets with fine grain control.

Scofield started the mid-tier bombing campaign by dropping more of the Massive Ordnance Air Burst bombs (MOABs) in an arc at the twelve-mile mark from the now-decimated Islamic front line. A ring of mushroom clouds, visible ten miles in either direction, would accompany the damage and death, spreading alarm through the Islamic ranks. He then ordered the individual air bosses to focus interdiction attacks on Templar armor killers. Next he drew a line on his virtual battlefield to show Air Force commanders where he wanted the next line of MOABs to land: an arc three miles closer than his last ring of death.

Scofield interrupted his deadly symphony just long enough to ask the GOS intelligence avatars to report on the effects of the mid-tier attacks. Considerable panic was spreading throughout this sector, many of the less informed and disciplined soldiers thinking they were under nuclear attack; but, by and large, the Muslim units were holding their ground and breaking up into smaller units to fight.

"How can anyone hold his ground under such brutality?" a frustrated Scofield asked his subordinates on the senior command network.

His reply came from Adm. Wainwright: "Our enemies are fighting men of faith. We have invaded and occupied their most holy ground and threatened them with its destruction. We should not wonder at their resolve. We may have to kill them all to leave this place in one piece."

"Do we have enough ordnance to kill them all?" Maria asked.

"No," said Adm. Franklin, meaning, *not without nuclear weapons.*

"Keep pounding the middle ranks until they break," Scofield commanded. "I'm going to take Miles Franks' regiment up the middle of the Islamic formations and split them in two. Col. Olsen will accompany me, and Gen. Marks will assume command for the remaining two Templar Regiments until I return. Gen. Marks, I want you to complete the slaughter of the enemy forces within three miles of the original front lines. This is where they have their last elite divisions. Cleanse the ground."

CHICAGO

"General, Col. Melissa Franks here. I have bad news. The FBI decided to contact the terrorists holding the bomb."

"Didn't Donner tell them to stand down?" Scofield asked.

"He discussed it with them, they told him they had sniper scopes trained on every terrorist so if anyone tried anything they would shoot and take them down before they could trigger the bomb. Donner gave them the go ahead," Franks said.

"What happened?"

"The bomb went off, it was in a warehouse near the downtown area."

"Shit! Shit…, damn, fuck!" Scofield's curses raised an octave on each successive word, after a short pause to regain his composure, he asked. "The yield?"

"Two hundred and fifty kilotons, about the same size as the New York bomb," Franks said.

"Gen. Marks, connect me to the president. Meanwhile, get the regiments positioned for their attacks and continue the air campaign without mercy."

RESPONSE

"**M**r. President."

"Gen. Scofield."

"Don't say it, I know, I screwed up, I gave the FBI more credit then they deserved." The president preempted the expected rebuke from his old commander.

"My Templars could have snuck in and severed the hands of the guy holding the trigger before anyone could have blinked." Scofield was incensed and the timing of this disaster could not have been worse. His frustration was palpable. "How do you want to respond to this latest attack?"

"My decision was wrong, and it has ramifications. If I don't respond aggressively, I think the hardliners will withdraw their support. They want me to nuke Iran."

"That action might not be containable." Scofield's mind was spinning, trying to figure a way out of this nightmarish scenario. "What do you want to do?"

"I think nuking Iran, wiping it from the face of the earth, might have a positive effect on your mission, a demonstration of our resolve and intent. It would also serve as punctuation for the little gifts you intend to leave when you finish your invasion. We didn't start this but we must finish this," President Donner posited. "What do you think?"

"Perhaps you are right. What we contemplate here is a great tragedy though, one I'd hoped to avoid. We counted the Israeli dead by the tens of thousands; the Iranian dead will be counted by the tens of millions, most of whom will be innocents."

"I know, but I am left cleaning up the millions of our dead here, they too were innocent," Donner lamented.

"Don't make this decision in anger, Mr. President, consider the impact and do it only if you truly believe it will be a net win in the long run."

"Roger that Mike, Donner out."

LIVE AND IN HIGH DEFINITION

The television feeds coming from the battlefront had gone from amazing to horrific. For the viewing audience, a good part of the world's population, it was unlike anything they had ever seen. Knowing that the feeds were *actual* combat was disconcerting enough, but knowing that the feeds were *live* was quite another thing. They had watched the battle at Bahrah and thought that it couldn't get any worse. But today, in the span of a few hours, they had seen unspeakable horrors; live, in high-definition color, and computer enhanced. They'd seen Gen. Scofield mutilate the terrorist's champion, and then the terrorist himself. But the scenes from the main battle were beyond violent. Most of the networks in the United States and Europe delayed the broadcasts, filtering out the most gruesome parts. Not C-Span; its feeds were live and unfiltered, with ratings off the charts.

It became a sickness. People couldn't stop watching.

Mouse had assigned a team of four program directors to manage the feeds and select the sources to transmit. They had developed programming avatars to automatically select video feeds based on specified criteria. The lead editor was spellbound in the process, absorbed in his macabre task, determined to elevate his morbid art.

Mark Osborne had been left behind to explain the technology and battle plans to the lucky fifteen embedded reporters assigned to Scofield's base camp. The reporters were new to nanotechnology, and understandably nervous as rounds impacted near the base camp. They assumed they were in mortal danger, but Mark Osborne, their colleague, remained unfazed.

"Mark, I understand how they're getting the feeds to us, but I don't get why the general wants us to see such grisly scenes of death and destruction," a *Newsweek* reporter commented.

"I don't know his motivation. You'll have to ask him."

"Why do you have a nano-armored suit?" another reporter asked, "Did they give you a helmet too?"

"They gave me a non-combatant version, so that I could traipse around and see the battle without getting injured. And, yes, I have a helmet."

"Can we get one?"

"No."

"What about the rumors of genetic enhancement?" one of the female reporters asked.

"I've no specific knowledge about that. You need to ask the general."

"We can't ask the general. He's fighting a war. So we're asking you," another reporter said.

"I have no specific knowledge," Mark reiterated. "I've learned a lot about nanotechnology."

"How can the Templars create this stuff so fast?"

Finally, a question he could answer. "Manufacturing as we know it may have changed forever. Dr. Mary Louise Monahan is Gen. Scofield's chief engineer. She has developed a nanobot lattice device that's, basically, a tiny robot brain with articulated arms that can be programmed to perform almost any function. She made these shelters that we are in, drawing their shape, pointing the Global Operating System at a clump of raw lattice nanobots, and telling the nanobots to self-organize into the thing she had designed. Once she had the structure set up, she could add whatever functional layers she wanted. Cloaking, armoring, air conditioning, *et cetera,* are applied, and the structure is ready to go. With a little imagination, a person at a computer can now design and build the most complex machines in the world, with no factories and no workers. All he needs are raw nanobots and a computer. And every device created is a fully functioning computer that can be integrated into what we call the GOS, to provide extra computing power."

"'What *we* call the GOS?'"

"What the Templars call it. Don't bust my balls. I don't have to be here babysitting you guys."

"Who supplies these nanobots?"

"Scofield Laboratories, the company where Dr. Monahan works."

"So you let your apostate general run your war to massacre the people of Islam *and* you also pay him to provide the means for that massacre. Is that right?" asked a dark, exotically beautiful female reporter from Al Jazeera.

"That's one way to look at it."

"How is it possible that you can send live pictures of the slaughter of innocent and peaceful Muslim people to the nonbelievers around the world? Do you think this is a video game? That the slaughter of my people is the next *Grand Theft Auto*?"

"I can't answer that. I'm not part of the Templar Division, and I don't make the decisions. I'm just here to answer the questions I can, because I've been living with the Templars for the last three months. You should direct your specific questions to the public affairs officer."

"They're apparently too busy killing Muslims to answer questions."

"Mark, this is Jennifer Singleton, remote from CNN headquarters in Atlanta. How are you today?"

"Fine, Jennifer, how can I help you?"

"We've heard rumors that Gen. Scofield has stolen Pakistan's nuclear weapons. Is that true?"

"Well, if he had, I don't think he would tell me that. I can't confirm it either way."

"He told you he intended to invade Mecca why not tell you about the nukes?"

"Don't have any idea, Jennifer, sorry."

Jennifer looked away from the monitor she was using to communicate with Mark Osborne half a world away, distracted by something. "Oh, my God, they just bombed Chicago."

IRAN NO MORE

"Adm. Franklin, this is President Donner, I have authenticated my order to attack Iran. Do you concur with that authentication?"

"I do, sir. I want to verify that you want the complete destruction of Iran, all city centers, military facilities and repositories, electrical infrastructure and peoples, is that correct, Mr. President?"

"Yes—proceed at your discretion."

"Very well, Franklin out." The admiral looked over at Wainwright to gauge his reaction. Wainwright's expression showed his revulsion at their orders. "Do we dare disobey Paul?"

"We have our duty," Adm. Wainwright said.

"We have a conscience and a soul too."

"There is that." Wainwright felt weak at the thought of what he had to do.

"Work up the attack plan. That will give us time to think and to contemplate the ramifications of what we are about to do." Franklin dismissed Wainwright to his task with that order. Wainwright pulled himself up out of his chair with great effort and dragged himself out of the room. He hoped he would have the stamina to complete his ghoulish task.

HUNTING THE GENERAL

There were four teams, each with the same mission: kill the apostate general. The Saudi prince had ordered them to avoid all combat and wait until the enemy weapons were no longer focused on the front lines. They were to emerge from their deep earthen bunkers on a singular mission: to seek out the general and kill him. They were told to look for a group of Templars surrounding a man who would seem distracted, even a little mad, waving his arms in the air. He would be the only one acting this way, with a large entourage. The teams were to search as stealthily as possible, communicate with each other, and coordinate their attacks.

They were issued semiautomatic, fifty-caliber sniper rifles acquired from Barrett Arms through a gun running American ex-patriot. They had explosive-head, armor-piecing bullets and a Russian device that would enable them to see the cloaked Templars. The bulky thermal-imaging device provided only fleeting images of the ghosts who roamed the battlefield. The snipers were the best Islam had. In the right hands, the rifles had a range of over a mile.

One of the teams had been killed in their bunker. A second had been located by a forward Templar scouting team and quickly dispatched. Of the two surviving teams, one was positioned in the steep hills just to the north of Ash Sharā'i`, the other in hills southwest.

Team One's leader reported a probable contact standing at the center of the Templar line, three hundred yards from the original Muslim line. Team Three's leader confirmed the sighting. They arranged to fire in tandem, Team One first, followed immediately by Team Three. By itself, the fifty-caliber explosive bullet would not defeat the general's armor, but three rounds impacting on him in rapid succession could defeat his suit's defenses.

They counted down. Three, two, one, fire!

The first bullet struck the side of Scofield's helmet, knocking him to the ground and disrupting his helmet's operation. The second round hit him square in the back, near his spine. Pain shot through his body.

Scofield's bodyguards pinned the general to the ground, surrounding him with their bodies and their custom nanoshields. With multiple fifty-caliber rounds striking their shields, Weaver knew instinctively that the enemy could see the general. He put out a frantic plea to the entire GOS-linked command structure to find and stop these assassins.

Firing a fifty-caliber rifle multiple times leaves an exhaust plume in front of even the most concealed sniper positions. Most snipers would have shot once and displaced, these snipers only had one chance—they had to stay in place and shoot it out. Raining down shot after shot on the huddled Templars, they saw two of his body guards die before they themselves succumbed to rocket attacks from two cloaked Apache helicopters, just one minute after firing their first shots.

Gen. Scofield was alive but badly banged up. His helmet fried, he opened his screen and saw the headless body of Capt. Weaver lying prostrate on the ground. He laid his hand on his comrade's chest and said a prayer for the dead, beseeching the angels to take his dear friend and protect him in his ascension.

Scofield rose, ready for battle, with no time for tears or remorse. He would remember and mourn Tim Weaver later. Taking the helmet of a bodyguard, he reconnected to the GOS and downloaded his specialized programs. Within a minute he was functioning again, directing the battle. The helmetless guard loaded the body of his revered captain into a waiting helicopter and returned to base camp to be refitted.

Watching the drama as it unfolded on the GOS-fed videos, Mouse and Maria breathed sighs of relief when they heard that Scofield was alive. Maria left her position on the left of the line and moved to the general's location. She would not allow herself to be separated from him again for the rest of the battle. He didn't have enough stars to make that order stick.

"Mouse, have you looked at my vitals?" asked Scofield, tormented by the pain from his two wounds.

"Not good, Mike. You could have a brain hemorrhage if we don't get you fixed up, and you have two fractured disks in your lumbar region."

"Can you patch me up remotely?"

"I've already injected targeted stem cells to the affected areas. If you give yourself time to heal, you should be good to go in a few hours."

"I don't have time to spare. We'll have to chance it."

"Can't you do that miracle thing on yourself?"

"Doesn't seem to work with me. I just heal fast."

"Do you want some pain meds?"

"I can't chance it, munchkin. I need to stay clearheaded."

"Aye, aye, sir."

Thirty minutes later, Mouse reported to the general how the snipers had been able to see them with a Russian thermal device. She would correct the problem and have the suits updated within a few hours. Whispering to herself, "I love you, Michael Scofield," she disconnected.

"KILL THEM, KILL THEM ALL"

Col. Miles Franks ordered his regiment to disengage with the enemy forces and assemble in the middle of the Islamic lines. He put all three battalions on line and awaited Gen. Scofield's arrival. With few surviving Islamic soldiers at this location, they faced little resistance and could rest and reflect on the morning's battle.

Some of the Knights wept uncontrollably, their suits reporting their psychological injury through the GOS system. Col. Franks visited the six men and two women most afflicted, who proved to be inconsolable in their grief and revulsion. They had seen too much bloodshed and could kill no more. They were used up. Franks felt pity for them—and even more for the men and women who remained, equally revolted, but soldiering on. He knew all too well that each would pay for their commitment later in life.

Knowing Mike Scofield's policy of serving only with volunteers, he gave each of the eight Knights the option of staying or leaving. All chose to leave. He thanked each for serving and doing their best, then relieved them and sent them back to base camp, to be returned to the ship. Their service would be honored, but they would never fight as Templars again.

When Scofield arrived, Franks told him about the eight psychological casualties and their relief from duty. Scofield said there would be more and that it was to be expected.

"You did the right thing, Col. Franks. We only serve with volunteers. When men and women have given their all and can give no more, they must be honored and released from their obligation. Now tell me how you're deployed."

"I have the regiment with three battalions on line, and all companies and platoons on line as well. The platoons are staggered into wedge formations, so we do have some depth. This will give us the maximum swath through the enemy, while making sure we cover the ground adequately. With twenty-two hundred men left, our regimental line stretches one mile wide. What're your orders, General?"

"We'll attack immediately and run though the Islamic line from south to north, then turn around and cut through them again and return to the division."

"Are we to take prisoners?"

"No. Kill all enemy troops and destroy all instruments of war. If an enemy tries to surrender, either ignore him and move on, or kill him, as your conscience dictates. There will be no reprisals for the killing. We've suspended our adherence to the Geneva Convention. This is a matter of personal conscience, not the law of war. Understood?"

"Yes, General. What will you do if you see a man trying to surrender?"

"I'll kill him."

"What about you, Maria?"

"I don't kill gratuitously, but I'll not judge others who do, not today."

Just as Col. Franks was about to signal the attack, a lone figure emerged from the distant Islamic lines. He was dressed in a black Bedouin robe that billowed in the wind. The robe was long and rich looking, giving the man an ethereal and threatening look. He was striding toward Scofield with purpose and without apparent fear. It would be impossible to find a figure more out of place on these killing fields.

"Don't touch him, he's mine!" Scofield ordered.

THE FATE OF PERSIA

"**M**r. President, we're ready to execute the plan of attack against Iran. Adm. Wainwright, who developed the plan, has conferred with Gen. Scofield who approved the attack plan, as have I. We want to make sure, before we proceed, that it's your intent to execute this plan with the certain knowledge that we're about to kill seventy-five million people. Are you sure this is what you want? What the American people want? And what is moral and just to avenge our dead?" Adm. Franklin, having reviewed the plan a few hours earlier, had gone into his private stateroom's head and threw up. He wasn't in favor of this action.

"While you were developing the plan, we convened Congress, and they voted to declare war against Iran. We then briefed Congress in secret session about our general plan of attack, and they voted their approval," President Donner said.

"The decision is yours, Mr. President. Please answer my question. Is this right and just?"

"It's necessary, Admiral, and you have been ordered to proceed."

"Shouldn't we at least ask for unconditional surrender before we annihilate the entire country?" Franklin asked.

"Who is available to respond? Scofield killed most of their leaders," Donner said.

"It's worth a try," Franklin pleaded.

"Very well, Admiral. We'll contact the Iranians through the Swiss embassy. I'll give them two hours to respond. Then, if they haven't replied, we'll proceed. Agreed?"

"Yes, sir, we'll be ready."

THE DUEL

The Saudi Prince moved purposefully toward Scofield without rushing. Scofield moved in front of his entourage about ten yards to meet him. They stopped ten feet from one another to assess each other.

"Good afternoon, General."

"And to you, Prince."

"Quite the show you put on today. Are you proud of your slaughter? Not much of a fair fight was it?" The prince was relaxed and confident.

"You orchestrated it; you should know," the general said. "I was happily relaxing at home in Virginia when this all started, playing with my granddaughter. We didn't start this war."

"Oh but you did, many centuries ago." He paused then taunted, "Shame about your family. General, I'm afraid I cannot save them."

The general's eyes narrowed at this threat, but he didn't respond. He prayed instead. *Please, Lord, give me the strength to die bravely this day, give me the strength of a thousand men that I might slay this evil prince. Take me, but only after I have freed the world of this menace.*

"Praying to Him won't save you today, General."

"I shall pray nonetheless."

"Prayer from a doubting Thomas like yourself isn't prayer…it's begging. Is that your whore behind you? She looks like she is ready to pounce. Does she do your fighting for you?"

Maria moved three feet toward the black-clad prince but the general signaled her to stop. "Maria will make her acquaintance known to you soon enough." The prince smiled at that comment. *Arrogant bastard.* Scofield thought.

"You don't look like a Saudi." Scofield taunted.

"I chose to look like myself today, as I looked when I was a alive." The prince replied.

"Let me guess, the 9th century, right?"

"You know your history, and your prophecies. Shall we begin, Michael Benjamin Scofield, General, United States Marine Corps?" The prince spat out the last sentence laced with venom.

"By all means."

The black prince attacked Scofield with a speed that shocked all of the Templar witnesses. Those Templars close enough to witness the confrontation began to move instinctively toward the general to come to his aid.

"No! This devil is mine! Stay back! That's an order."

Scofield dodged the prince's sword thrust and turned to swing his sword viciously at the prince's head. The prince ducked and came back to Scofield with an equally vicious counter thrust.

The fight went on with thrust against counter thrust for what seemed like hours to the hundreds of Templars watching. Most of the other Templar units were sent a video feed of the battle, which caused most ground fighting to come to a halt along the entire expanse of the battlefield. Mary Louise, ensconced at her terminal, was transfixed, praying for the general to prevail.

The Templars sensed that Scofield, wounded though he was by the earlier assassins, was starting to get an upper hand in the fight. His thrusts were increasing in violence, and his speed and accuracy nearly eviscerated the prince on numerous occasions. Scofield was pissed, spitting malice and closing for the kill.

It was then that the prince struck a blow that cut Scofield's nanosword in half. The prince then blocked Scofield's fist, turned, and thrust his sword through Scofield's midsection. It all happened in the blink of an eye.

They faced each other, Scofield impaled on the prince's sword—the sword extending through the general's back by six inches, trailing blood and bits of his entrails. The black prince then raised his sword, cutting up and through Scofield's lower torso to his rib cage. The general groaned pitifully. As he began to turn the sword to finish his kill, the prince saw Maria closing the distance between them in a flash of scarlet plume. He heard her war cry and sensed her projected rage.

Scofield watched helplessly, feeling the searing pain in his side as Maria chopped down on the prince in a blow so violent that it would have sliced a tank in half. The prince's robes billowed as they were sliced in two.

The black prince had disappeared.

THE WOUNDED GENERAL

"Oh my God, Michael!" Prepared as she was for this horror, she was still overwhelmed by the sight of it. "Munchkin, vector a bird here, tell the trauma surgeons to prep the OR."

"Did that when I first saw the Prince," Mary Louise said. "It will land in thirty seconds. Tell the General to hold on."

"Munchkin says to hold on, General. I'm ordering you to."

"I'm dying, Maria. You have to take charge of the battle. I've already briefed Gen. Marks; he supports my decision. You're hereby given the temporary battlefield commission of major general." Scofield moaned as he bled profusely from his wound. The sword remained in his body. "I don't have much time. Swear to me you will finish this."

"I have already given my oath. I reaffirm it here, my love. Go with God. Find peace."

As Maria said these words, the medical team from the chopper reached their position. Scofield lost consciousness as they loaded him on the bird.

Maria's tears burned her eyes as she rose. Her body shook with rage. As she calmed herself for the task ahead, she tasted blood. She had bit through her inner lip. *Good,* she thought, *I want the taste of blood on my lips for the rest of this battle.*

"How did the Prince's sword penetrate the General's nanosuit, Munchkin?" Maria asked.

"Don't know. Black magic or damn good technology. I'll know when I see the sword," Mary Louise said. "Will he live, Maria?"

"Only if God wills it. No man could survive that wound without divine intervention."

"Where did he go?" Mary Louise asked about the black prince.

"I don't know. I was right on him; no technology could do that."

THE END OF PERSIA

"**D**onner sent word a few minutes ago that the surviving leadership in Iran told him to go fuck himself. We've done all we could do to stop this; now there is just duty or mutiny. What say you, Adm. Wainwright?" Franklin and Wainwright were alone in the command center after Franklin had ushered out the operations crew.

"If we refuse to launch, then they will launch ICBMs from the US. That's much riskier because the intended destination of the missiles might be misinterpreted. It's better that we launch with local assets. They will be in flight for less than a few minutes…and their intended impact known before anyone can question what's happening. We launch the attack, Mike. We do our duty and pray for the innocent we're about to slaughter," Wainwright said.

"May God forgive us Paul. Launch the attack when ready."

The attack was multidimensional in nature. A mixture of airburst, high-yield thermonuclear bombs carried by cruise missiles against hardened sights and various military and civilian targets coupled with neutron bombs in less-populated regions. To cap off the attack, Wainwright launched tactical EMP weapons to cripple anything left untouched.

The attack took two hours. When it was over, the nation of Iran and the Persian peoples were nearly wiped out.

"Let's hope the fourth bomb was in Iran when we attacked." Franklin said quietly, leaning over to Wainwright.

"We can only hope," Wainwright said.

"Still, let's increase satellite and airborne recognizance on the eastern part of our area of operations. Pay close attention to the Saudi airfields. They have some deep bunkers from the first gulf war that may have survived our bombing runs," Franklin said.

"Roger that."

AFTERMATH

"Gen. Olsen, the troops are rattled by Gen. Scofield's death and the news about Chicago and Iran." Gen. Marks had joined Maria on the battlefield, leaving the headquarters operations to Col. Melissa Franks.

"He's not dead, not yet; they will tell me the moment he passes. *If* he passes," Maria said defensively. "Munchkin, you listening?"

"Yeah."

"Open a channel to all Templars."

"Christian warriors of the Templar Division. This is Maj. Gen. Maria Olsen, acting commander. As you know, Gen. Scofield was wounded in battle hours ago. He has been transported to a hospital ship and is in surgery for his wounds as we speak. The general ordered me to lead you in the final stages of this battle. It's my honor to do so. The Navy has destroyed Iran. Now we'll bloody this field to avenge our fallen general and to avenge our countrymen murdered in New York, Washington, and now Chicago. Strength and Honor," Maria opened the division's microphones to hear the replies of her troops.

"And a good death!" resounded the surviving Templars.

They won't falter or fail...they're all lions today. Let's finish this! Maria thought. "Olsen out."

THE FINAL ATTACK

Col. Franks sent his regiment into the attack on Maria's order. In the past, a regiment on line, spread out a mile wide and running, would have been impossible to manage. With the GOS, and 360-degree vision, it was easy for each Templar to know where he was in the formation and where the other units were. Each man could pace himself, and commanders could easily correct the line if needed.

The regiment encountered scarce resistance for almost a mile into the Islamic lines. Then they came upon a mechanized regiment trying to reorganize after a devastating air attack. The Templars hit the regiment fully cloaked and cut through the men, machines, and supplies in short order. The Muslim regiment, still reeling from the earlier attack, dropped their weapons and equipment and ran for their lives—just as Scofield had hoped. Miles's regiment slaughtered them as they fled.

The Templars typically used their chemical bullets to kill men and their swords to permanently disable vehicles, tanks, or artillery. The work was long, arduous, and gruesome. There seemed no end to the Islamic armies. The Muslims died by the thousands, mile after mile.

THE LAST BOMB

"Adm. Wainwright!" One of the command center operators waved urgently toward the admiral.

"Yes, what is it?"

"Five aircraft have taken off from a civilian airbase east of Riyadh."

"Identity?"

"F-16s, probably Saudi, traveling at ground level."

"Distance to the battlefield?"

"Less than five hundred miles now and closing."

"Speed?"

"Mach-2."

"At that height?"

"Don't think they care about the damage, sir."

"Time till they enter the Templar Division Theater of Operations?"

"Fifteen to twenty minutes, depending on their vector and any evasive action they take."

"What intercept assets do we have to deploy?"

"We have our combat air patrol on station at fifteen thousand feet, four aircraft ready. We also have ground-to-air defense assets surrounding the Templar base camp. They can reach out and hit the attackers if they approach, but they don't cover much of the main Templar battlefield. The Templars have some handheld, ground-to-air defense capability, but that won't help if these are the nuke aircraft you were worrying about."

"Thank you, Chief. Send all available assets, including attack aircraft with any air-to-air missiles on their wings, to intercept the five planes. Give me constant updates as to the progress of the intercept."

"Aye aye, sir."

GET YOUR HEADS DOWN

Wainwright opened a line to Maj. Gen. Olsen. "General, we have a problem."

"What is it?"

"I think we found your nuke. It's probably on one of five Saudi F-16s headed right for us from Riyadh."

"What're your chances of shooting them down, Admiral?"

"Very good, but even a proximity explosion would be bad. They're about fifteen minutes out, moving at mach-2." Wainwright said.

"The Templar division is highly dispersed over a thirty-mile range of operations. That's good, but I'll inform the unit commanders to prepare to dig in all Templar Knights starting ten minutes from now just in case. The nanoarmor will protect us if we bury ourselves in the sand from any explosion about two miles out. It's never been tested, but I think that's a good operating distance. They can't kill many of us," Maria said. *Damn it*, she cursed to herself.

"What if they hit the base camp, Maria? Can Mouse's nanostructures withstand a direct hit?" Wainwright asked.

"No. Do you think they would explode a nuke so close to the Holy Mosque? That was one of the reasons we chose the location for the base camp," Maria asked.

"Do I think the Saudis would? No, I don't. But I think the black prince would."

"Yeah, maybe…and then blame us for it." Maria said.

"What do you want me to do about base camp?" Wainwright asked.

"Vector all available choppers that can reach base camp and depart by the deadline to evacuate key personnel. I'll notify Col. Franks and give her instructions. Do you have any idea how many people we can move in that timeframe?" Maria had one thought, *Mary Louise!*

"Don't know but I'll get you that number as soon as I can," Wainwright said.

"Just tell Melissa Franks. I'll give her the priority list, and she can manage it."

"Let me guess who is first on that list."

"Damn right, Admiral."

"Olsen out."

"Good luck, Maria. Dig deep and keep your heads down."

Maria then opened a priority communications line to Melissa Franks and gave her the rundown of the nuclear threat.

"Get Mary Louise and her engineers and the reporters out on the first two choppers, in that order. Then get others out as soon as possible as you see fit based on their needed skills in battle support and operations. If Mary Louise refuses then tie her up and throw her on the bird."

"Will do, General."

"And I want you to get on the last chopper. You're too important for us to lose. Move the operations center back to the ship."

"I won't leave my troops, General. Won't do it."

Neither would I, Melissa... God bless you and good luck, Maria thought. She didn't bother to argue. Maria nodded her head to signal that she understood. "Good luck, old friend."

"Tell my husband to dig deep. Tell him I love him."

"I will."

INCOMING

There was nothing more to do but wait. All Templars were dug into the sand or ground wherever they happened to be. Gen. Olsen had ordered each Knight to get at least three feet underground and to bury him or herself with dirt, sand, or debris from the battlefield. Their communications were still operating so each Knight remained connected to the outside world. Given the suits they had to wear, there weren't many claustrophobic Templars, but being buried alive while awaiting a nuclear bomb could be disconcerting to the bravest souls. Everyone was grateful for the communication links.

Adm. Wainwright, anticipating the anxiety of the division, was sending real-time tactical updates to the division's troops.

"Admiral, we've intercepted three of the F-16s on approach. We have two more to go. There were no secondary explosions, so either we haven't hit the aircraft with the bomb, or it was destroyed. We don't have the ability to test for radiation signatures at the crash sites—we're working on that. We should assume the bomb is still aloft," Wainwright's chief of operations said.

"That would be rotten luck. How close are we to the other two?" Wainwright asked.

"Missiles have been launched at number four. Number five has slowed and is using map-of-the-earth in the low hills between Mecca and Riyadh."

Both men were frozen in place, waiting for the results to come in.

"Got him!" yelled one of the operators from his console in the front of the operations center. "Number four is dead, no secondary explosion reported."

"Hot damn!" said Wainwright.

"Missile launch detected."

Wainwright looked to see which operator had said it. He wasn't sure. Everyone was absorbed with his work.

"Detonation one-half mile east of the Divisional Base Camp," the same voice reported.

"God damn it! Who said that? What detonated?" Wainwright shouted.

Chief Leiter turned to the admiral to report. "Nuclear detonation over the Templar Division Headquarters at an estimated height of one thousand feet. Estimated yield is twenty-five kilotons. We've lost communications with Col. Melissa Franks."

"Did it come from the last aircraft we were tracking?" Wainwright asked.

"No, sir, number five is dead; they must have been decoys," another operator intoned.

"Twenty-five kilotons…that's tactical nuke range. That bomb couldn't have been the last Iranian bomb," Wainwright mused.

"I think we just got a little payback from our Chinese friends."

END OF THE LINE

"We can't be sure, Miles, but Melissa may have died in the explosion. We lost communications with her and the entire base camp after the explosion. We've sent Templars into the blast zone, and they describe the place as being utterly destroyed. So far we've found no survivors." Maria was briefing Col. Miles Frank. "Do you wish to withdraw from the battle? Gen. Marks and I can complete this final sweep."

"No, no. Thank you, Maria. I'll mourn my wife later. She'd kick my ass when I see her in heaven if I left my command." Miles was near tears, his wife was irreplaceable to him, but he would do what they had always done together: he would do his duty and honor her death. "How many died at the base camp?"

"We lost nearly a thousand. Over four hundred Templars, and many Army and Navy support personnel," Maria said.

"Mouse?" Miles asked.

Maria chuckled, "She refused to leave, so Melissa tied her up and carried her to the chopper. When she kicked and bit your wife, Melissa put her over her knee and spanked her…I shit you not." Miles and Maria shared a laugh, imagining the scene.

"She was one of a kind, your wife."

"Yeah—they broke the mold."

"Finish this then?" Maria asked.

"Payback." Col. Miles Franks saluted and turned to go back to his head-quarters team.

The attack by Miles's regiment had been halted for the nuclear emergency just before reaching their initial objective, which was the end of the Islamic lines some thirty miles north of the initial line of departure. The regiment covered the last two miles quickly and without much resistance.

After reaching their objective, Gen. Olsen ordered Col. Franks to move the regiment two miles to the west and prepare to reverse the direction of march to attack back to the divisional position. The men were tired and distressed by the slaughter required of them, but they soldiered on. Consulting the GOS intelligence map for a virtual flyover of the battlefield, Maria was amazed at what she saw. The battlefield, fifty miles wide in places, was ablaze with burning vehicles and other debris. Men milled about dazed and unorganized, mostly unarmed, searching for a way to survive the attacks from the murdering ghosts.

The regiment began its homeward movement, catching men who had fled west from the earlier attack. The battle was no longer two-sided. The enemy was a disorganized rabble fleeing for their lives. The regiment cut through them mercilessly.

At the halfway point on their return trip, the regiment reached a tiny outpost called At Tarfa`. There were no Islamic soldiers left to kill. Adm. Franklin's aircraft had already eliminated the forces stationed there. The women of the town knelt by the corpses, overcome with grief. They beat their chests and wailed. The Templars moved through the scene like ghosts, leaving the women to their grief.

Maria looked at the women tending to these dead strangers, touched by the scene of despair arrayed before her. *We've done enough, no more killing today.*

WITHDRAWAL

A t the peak of the battle, when the remnant Islamic armies were collapsing in disarray and despair, acting Maj. Gen. Maria Olsen ordered her surviving forces to withdraw to the port city of Ash Shaiba for extraction. The surviving support units were extracted first, over the course of many hours. The Templar division extracted by foot along Highway Five, killing any armed soldiers along their path until they reached the area south of Mecca. The disorganized remnants of the Islamic armies sensed a change in the tempo of battle, but they were unable to respond. When the Templars passed Mecca, the battlefields went silent, and the Muslim survivors were left to ponder their fates and bury their dead. It was as if their invisible tormentors had disappeared from the face of the earth.

When the Templars finally embarked on their ships, a lone helicopter landed at the foot of the Kaaba in Mecca. Maria stepped out with Tommy Marks and their bodyguards. The few Muslims on hand were mostly religious men beseeching God's mercy and forgiveness.

Maria could see damage from the nuclear explosion that had detonated a little over ten miles away, but the Holy Mosque and the Kaaba looked to be untouched.

A flying camera showed the general standing alone in front of the sacred Kaaba. The video was fed live to the world's news organizations with the headline: *Maj. Gen. Maria Olsen declares end to hostilities, will address Islamic nations.*

"The United States of America ends its hostilities against the combined Islamic nations, effective immediately," Olsen proclaimed. "Our thirst for vengeance has been sated. We seek no further retribution for our five million murdered countrymen.

"We expect the Islamic forces to withdraw from their positions surrounding Israel immediately. Israel will choose its new borders with its Islamic neighbors, and all Muslims, military and civilian, shall vacate Israeli lands for all time. Henceforth, any attack against Israel shall be an attack against the United States.

"We seek peace with a new Islamic nation. In case that's not possible, we agree unilaterally not to aggress against Islam unless we're first aggressed against. Know this, however: if there is *any* aggression from Islamic forces or your surrogate terrorist organizations, we will lay waste to Muslim lands and burn its surface with nuclear fire.

"Look to the north," Olsen said, pointing to a bright light a half-mile north of the Kaaba. Another light could be faintly seen due west, from the ruins of Jeddah.

"These lights shall remain on, and the weapons within their enclosures shall remain armed, until there is a demonstrable peace between our peoples. The lights you see emanate from nano-armored spheres that house ten-megaton nuclear warheads. These lights are now appearing in Cairo, Damascus, Jakarta, Islamabad, and all the great cities of the Islamic nation. These weapons will be detonated instantaneously by automated control if we're attacked.

"To the faithful of Islam, seek Him, and seek peace with the other peoples of Earth. Don't force your beliefs upon free peoples. Let them follow you by your example, or not at all. I honor your brave dead. I wish them peace as they move on to seek God in His Kingdom. They fought bravely, as true believers, and against impossible odds that no man could have overcome. Honor them."

THE LAST GOODBYE

The Blackhawk started its engines, and the general's party boarded. It lifted off, cloaked, and disappeared. Maria asked the pilot to fly over Jeddah, the once-beautiful city she and Scofield had destroyed. He had them put down on the outskirts of town in the assembly area where they had stepped off to attack Mecca.

"The general just died, Maria. He never regained consciousness. I'm very sorry."

"He passed at the moment we landed." Maria responded, her tone indecipherable.

Marks eyed Maria.

Invisible to the survivors of the battle, Maria and Marks surveyed the damage before them. They stood motionless, witness to the carnage surrounding them. Death and despair for the vanquished Islamic armies, for Saudi Arabia, Iran, and the Muslim world.

"Are we good or evil?" Maria asked.

"Evil feels no remorse; we do," Tommy Marks said.

"I understand him now…" Maria looked into the distance, weary to the bone.

Both warriors stood in silence, unable to break the melancholy.

Coming Soon: The Sequel to "The Lion's Prophecy"

Comes Now the Darkness...

About The Author:

Michael Gaddis spent ten years as a Marine Corps officer. He served with the infantry in the First and Second Battalions, Third Marine Regiment out of Kaneohe Bay, Hawaii, where he participated in multiple West Pacific cruises with the Pacific Fleet. He served as a platoon commander, company executive officer, and company commander with the battalions. After his tour with the grunts, he added a computer-and-data-communications military operational specialty (MOS) to his skills by serving with the Marine Corps Data Processing center in Kansas City and the Defense Communications Agency. Along the way, he picked up his bachelor's and master's degrees in Computer Science.

After leaving the Marine Corps in 1988, he joined the Applied Research Laboratory at Washington University in St. Louis, serving as its associate director until 1993. The laboratory specialized in commercializing faculty research and managed to license and commercialize two leading-edge, high-speed communications switches to technology companies during his tenure.

After his stint at the university, Michael Gaddis helped create three successive technology companies concentrated on Internet technology and services. He would serve in the capacity as director, executive vice president, chief technical officer, and chief executive officer during his "entrepreneurial" years.

He retired in 2001 at the age of forty-five to pursue fun things like writing software and novels such as *The Lion's Prophecy*.

www.ingramcontent.com/pod-product-compliance
Lightning Source LLC
Chambersburg PA
CBHW070918180626
46817CB00003B/1119